DEEP TIME RECKONING

One Planet
Sikina Jinnah and Simon Nicholson, series editors

Peter Dauvergne, *AI in the Wild: Sustainability in the Age of Artificial Intelligence*
Vincent Ialenti, *Deep Time Reckoning: How Future Thinking Can Help Earth Now*

DEEP TIME RECKONING

HOW FUTURE THINKING CAN HELP EARTH NOW

VINCENT IALENTI

THE MIT PRESS   CAMBRIDGE, MASSACHUSETTS   LONDON, ENGLAND

This book was set in Stone Serif, Stone Sans and Avenir by Westchester Publishing Services. Printed and bound in the United States of America.

Library of Congress Cataloging-in-Publication Data

Names: Ialenti, Vincent, author.
Title: Deep time reckoning : how future thinking can help Earth now / Vincent Ialenti.
Description: Cambridge, Massachusetts : The MIT Press, [2020] | Series: One planet | Includes bibliographical references and index.
Identifiers: LCCN 2019053898 | ISBN 9780262539265 (paperback)
Subjects: LCSH: Radioactive waste disposal in the ground—Finland. | Radioactive waste repositories—Finland. | Environmental policy—Finland. | Nature—Effect of human beings on—Forecasting. | Global environmental change—Forecasting. | Human ecology. | Social planning. | Geology, Stratigraphic—Anthropocene. | Time—Social aspects. | Finland—Environmental conditions—21st century.
Classification: LCC TD898.13.F5 I25 2020 | DDC 363.72/89094897—dc23
LC record available at https://lccn.loc.gov/2019053898

10 9 8 7 6 5 4 3

CONTENTS

SERIES FOREWORD

This is at once an odd and exhilarating time to be alive. Our species, *Homo sapiens*, has had roughly 350,000 years on the planet. For most of that time our ancestors barely registered as a quiet voice in a teeming chorus. No more. Now, a human cacophony threatens the ecological foundations upon which all life rests, even as technological wonders point the way toward accelerating expansion. We find ourselves at a moment of reckoning. The next handful of decades will determine whether humanity has the capacity, will, and wisdom to manufacture forms of collective life compatible with long-term ecological realities, or whether, instead, there is an expiration date on the grand human experiment.

The One Planet book series has been created to showcase insightful, hope-fueled accounts of the planetary condition and the social and political features upon which that condition now depends. Most environmental books are shackled by a pessimistic reading of the present moment or by academic conventions that stifle a writer's voice. We have asked One Planet authors to produce a different kind of scholarship. This series is designed to give established and emerging authors a chance to put their best, most astute ideas on display. These are works crafted to show a new path through the complex and overwhelming subject matters that characterize life on our New Earth.

The books in this series are not formulaic. Nor are they Pollyannaish. The hope we have asked for from our authors comes not from overly optimistic accounts of ways forward, but rather from hard-headed and clear-eyed accounts of the actions we need to take in the face of some-times overwhelming odds. One Planet books are unified by deep schol-arly engagement brought to life through vivid writing by authors freed to write from the heart.

Thanks to our friends at the MIT Press, especially to Beth Clevenger, for guiding the One Planet series into existence, and to the contributing authors for their extraordinary work. The authors, the Press, and we, the series editors, invite engagement. The best books do more than convey interesting ideas: they spark interesting conversations. Please write to us to let us know how you are using One Planet books or to tell us about the kinds of themes you would like to see the series address.

Finally, our thanks to you for picking up and diving into this book. We hope that you find it a useful addition to your own thinking about life on our One Planet.

Sikina Jinnah and Simon Nicholson

FOREWORD

Ei ole mitään niin paljon kuin aikaa.
Nothing is so plentiful as time.
—Finnish proverb

How far have you traveled in time today? Your focus has likely been on the demands of the day, perhaps a deadline looming next week, a trip planned for a few months from now. Maybe a conversation sent your thoughts back to events that happened some years ago, or a Proustian aroma returned you momentarily to childhood. But has your mind ventured forward to the next decade, century, or millennium? Thinking about times to come—in anything more than the most generalized way—is hard work, both intellectually and emotionally. The future is an elusive abstraction we almost don't believe in; looking straight at it requires us to confront our fears and anxieties, the possibility of misfortune, and the fact of our mortality.

In *Deep Time Reckoning*, Vincent Ialenti presents an anthropological investigation of a contemporary subculture in which peering many millennia into the future is an everyday activity: the Finnish "Safety Case" project, charged with imagining how geological, hydrological, ecological, and social factors might interact over tens of thousands of years in the area around an underground nuclear waste repository on the island of Olkiluoto.

In the manner of other ethnographers, Ialenti spent almost three years doing fieldwork among those working on the Safety Case, observing and interviewing them, documenting their individual and collective habits, and abstracting from all of his data a set of practical strategies for envisioning possible futures not for the sake of the Safety Case workers or edification of other ethnographers but for the rest of us mired in the present. Ialenti calls these strategies "reckonings," a creative appropriation of a rather old-fashioned word that bears multiple meanings—calculation, struggle, verdict. His hope is that the techniques the Safety Case workers use to reckon with time may serve as templates for all humankind at this moment in history when long-term thinking is urgently needed.

Some of the Safety Case's ways of visualizing the future are temporal inversions of methods that geologists use to reconstruct the distant past. For example, to interpret an ancient volcanic rock with a particular crystal texture, a geologist looks for modern analogues produced by active volcanoes. Similarly, many of the Safety Case experts develop future scenarios by studying past and present analogues—a Mesozoic copper deposit in mudstone in Devon becomes a stand-in for copper canisters to be buried in clay at the repository site; the behavior of glaciers in modern Greenland provides insights into what Olkiluoto might experience in a future ice age. Like geologists, members of the Safety Case team habitually and casually zoom in and out of timescales.

Ialenti emphasizes that the Safety Case approach is methodologically omnivorous, combining analogue studies with a web of interconnected quantitative models to create a suite of scenarios for "far future Finlands." The Safety Case embraces "strategic redundancy"—integrating the logic of different fields to ensure that any one potentially inaccurate assumption does not invalidate the entire range of projections. This rigorous, multiscale interdisciplinarity is a refreshing counterpoint to the myopic, reductionist thinking that has produced many of the intractable environmental problems of the Anthropocene.

There is some irony in studying Finns as exemplars of future thinkers: as Ialenti points out, the Finnish language has no future tense. Instead, either present tense or conditional mode verbs are used, which seems a rather oblique way of speaking of times to come. But this linguistic treatment of the future may reflect a deep wisdom in Finnish culture that informs the philosophy of the Safety Case. Making declarative pronouncements about the future is imprudent; the best that can be done is to envisage a spectrum of possible futures and develop a sense for how likely each is to unfold.

Ialenti notes that another peculiarity of Finnish culture is the remarkably widespread trust in technical experts to make judicious decisions for the public good. Elsewhere in the world, a combination of class-based resentment, anti-intellectual populist political rhetoric, and internet-related dilution of authority have led to what Ialenti calls the deflation of expertise. In a profoundly hypocritical contradiction, the public that eagerly consumes new technologies increasingly rejects the results of science. Taken to its extreme, the deflation of expertise is nihilistic, the end of two centuries of Enlightenment belief in the power of rational thought to elevate humanity. We can only hope that outside Finland, distrust of all claims to specialized knowledge will soon be reversed, before lasting damage is done to the architecture of civilization.

While most of the lessons from the Safety Case are positive, some are cautionary tales. Ialenti dedicates an entire chapter to how the project nearly collapsed in 2005 when a brilliant but abrasive key expert in the project died unexpectedly, leaving cryptic records and simmering resentments with long half-lives. This story is a reminder of the tension between individual careers and collective work, the challenge of creating intellectual infrastructures that outlast any particular life span.

Ialenti acknowledges that implementing the deep time reckoning practices of the Safety Case in other contexts and cultures will not be easy. He suggests imaginative exercises—"calisthenics for the mind"—that may help us build our capacity for long-term thinking. His self-described radical optimism and dedication to intellectual resistance are restorative antidotes to the paralyzing pessimism so many of us feel when we dare to think about the future.

But if we don't learn soon to reckon with time—to see ourselves in proper temporal proportion, to prepare intelligently for our impending trip to the future—our time of reckoning will come. As the Finnish writer Henrik Tikkanen (1924–1984) wrote, "Because we don't think about future generations, they will never forget us."

Marcia Bjornerud
Lawrence University
October 2019

PREFACE

This book is a response to two overlapping crises—one ecological, the other intellectual. The first is the *Anthropocene*: a name proposed for a troubled time in Earth's history ushered in by human transformations of our planet's climate, erosion patterns, extinctions, atmosphere, rock record, and more. The second is the *deflation of expertise*: what happens when political power is commonly gained through populist mockery of expert authority, experts' voices are too often drowned out by the noisy clamors of knee-jerk tweets and self-published blogs, and adepts' inquisitive spirits are frequently dulled by new corporate-managerial reforms and constrained by bureaucratic protocols.

The first crisis, I will argue, challenges entire populations to reimagine their ways of thinking, acting, and relating to better sync with the Earth's environment's radical long term. It calls us to become more skilled *deep time reckoners*. This means repositioning our lives, ideas, and societies within wider time spans—better squaring our drives for short-term gains with the long-term well-being of our species. Yet the second crisis creates new obstacles to this. Today, it is increasingly difficult to put forth bold, evidence-driven, intellectually ambitious visions of the Earth's future, given widespread skepticism toward technocratic knowledge, liberal arts education, scientific research on the environment, and even the very possibility of there being verifiable facts, truth, or a single shared reality.

With these hurdles in view, this book argues that hearing out the world's most long-sighted experts should be step one in a fascinating, yet gravely important, trek toward widening our intellects' timescales in times of ecological downturn. This means opening our ears to astrophysicists, geologists, historians, cosmologists, anthropologists, evolutionary biologists, archivists, theologians, paleontologists, nuclear waste scientists, climatologists, philosophers, archaeologists, and other long-term thinkers. A bit like Sun Tzu's fifth-century BCE treatise *The Art of War*—a practical guide for cultivating a military general's mentality—*Deep Time Reckoning* is a practical guide in the art of long-termism.

I write this guide as a cultural anthropologist. From 2012 to 2014, I spent thirty-two months living and conducting fieldwork in Finland. I was doing research abroad for a PhD program at Cornell University. Over these months, I recorded many interviews among experts who used the toolkits of laboratory science, government regulation, computer modeling, geotechnical engineering, corporate planning, and office life to make ambitious forecasts of far-future worlds. Most of my informants were involved with or had something to say about Finland's nuclear waste repository project at Olkiluoto. Their goal was to permanently bury radioactive waste from the country's nuclear power plants. Striving to contain multimillennial half-lives, these experts reckoned with deep time. They forecasted geological, hydrological, and ecological events that could occur over the coming tens of thousands, hundreds of thousands, or even millions of years. Like other anthropologists doing years of fieldwork in cultures unlike their own, my mission was to uncover kernels of wisdom that could, hopefully, widen people's worldviews in my own or other societies. This book is the outcome of my journey. Its aim is simple: to glean, from Finland's nuclear waste experts, long-termist sensibilities that any of us can adopt to help resist today's dual crises of ecology and expertise.

Deep Time Reckoning is what anthropologists call an ethnography. It is a description and analysis of my fieldwork findings and interpretations. As an ethnography, though, it is unusual. I have not written it mainly for fellow social scientists, humanities scholars, or nuclear researchers. This book is for the educated public, expert and lay alike. It requires no prior knowledge of anthropology or nuclear issues—just a little patience in following me and my informants into far-future Finlands. We all live in times shaken by populist unrest, fractious social division, out-of-touch elites, and unraveling trust between experts and laypeople. As I see it, the last thing these unsettled times need is yet another alienating academic tome, hiding its best ideas behind impenetrable jargon. What's needed, rather, are more stable bridges between the social and natural sciences, experts and nonexperts, and diverse cultures inhabiting a damaged planet that, for many, feels less and less like a friendly global village each day.

So, *Deep Time Reckoning* is about more than just Finland's nuclear waste program. It aims to provide food for thought anywhere that taking a long view becomes useful. This could include projects of biodiversity conservation, information archiving, climate change mitigation, city planning, natural resource extraction, infrastructure security, digital technology obsolescence, landscape architecture, human mobility planning, land management, and beyond. After all, as the Anthropocene and the deflation of expertise take hold, we all have something to teach and something to learn. We all must embark, each in our own ways, on personal journeys of long-termist learning. Along the way, we must never fool ourselves into thinking our intellectual growth is complete.

Fortunately, my anthropological fieldwork has left me optimistic that human hearts, minds, and technologies can grow to secure better ecological tomorrows. My optimism, however, remains guarded. Placing an abiding-but-measured faith in science, technology, and intellectual inquiry is vital, but this faith must always be reinforced with humility, hard work, strong oversight, ethical self-examination, restraint, and self-critique. The same goes for our abilities to reckon with deep geological timescales. It would be more than naïve to assume that Finland's nuclear waste experts painted perfect portraits of worlds for millennia to come. They did not. In fact, many of them emphasized that their own computer models of the future, however sophisticated, were but highly educated guesses. They saw them as merely the most credible future-gazing strategies they could concoct using the best science and technology available to them at the time.

Yet it would also be naïve to deny that these experts have made, or can make, real progress in refining their long-termist techniques. Undeterred

by the deflation of expertise, this book is an exercise in adopting my informants' faith that a highly disciplined, tightly organized, well-trained,
adequately funded group of experts can, in fact, inch us closer to understanding futures near and deep: our most crucial Anthropocene task. It
is, in other words, an anthropological thought experiment in mimicking
their worldviews, and then expressing them in print. This book's contention is that embarking on imperfect projects to envisage far futures is,
ultimately, far more enlightening than giving up or never embarking
in the first place. In this spirit of adventurous learning, we can practice
stretching our minds across time.

ACKNOWLEDGMENTS

Deep Time Reckoning is a passion project that has consumed me for over
twelve years. I began framing it as an undergraduate major in philosophy,
politics, and law, taking a graduate-level anthropology course in ethnographic analysis at the State University of New York at Binghamton. I
continued developing it in a master's program in law, anthropology, and
society at the London School of Economics, a PhD program in sociocul-
tural anthropology at Cornell University, then as a postdoctoral fellow
and now assistant research professor at George Washington University's
Elliott School of International Affairs. I am grateful that, along the way,
my research has been supported by a US National Science Foundation fellowship (GRFP 2011129751), a Mellon Fellowship from Cornell's Society
for the Humanities, a Nuclear Waste Solutions grant from the John D. and
Catherine T. MacArthur Foundation, plus a generous package of teaching
assistantship, research assistantship, and fellowship support from Cornell's Anthropology Department.

I thank the MIT Press's anonymous reviewers for their wise comments
on earlier drafts of this manuscript, plus Beth Clevenger, Judy Feldmann,
Sikina Jinnah, and Simon Nicholson for their excellent editorial work. I
thank my doctoral committee—Annelise Riles, Hiro Miyazaki, and Mike
Lynch—for their tremendous patience and invaluable advice. I thank Alli-
son Macfarlane and George Washington University's Institute for International Science & Technology Policy for providing me with an enriching
academic home base for over three years. I thank the students who took

my fall 2016 seminar, Nuclear Imagination: Technologies & Worlds, for their stimulating questions. I thank the Meridian 180 think tank community for the vibrant conversations we have had together since 2011. I thank Cornell's Society for the Humanities' 2015–2016 fellows for introducing me to so many thought-provoking ways of thinking about time. Chapter 4 of this book is an extended version of an article I wrote titled "Death & Succession among Finland's Nuclear Waste Experts" (*Physics Today* 70, no. 10 [2017]: 48), adapted with permission from the American Institute of Physics. I thank *Physics Today*'s editors and reviewers for their helpful feedback.

Over the years, many colleagues, friends, mentors, and acquaintances have shaped my thinking through fascinating conversations, intense debates, and constructive comments on past drafts. To name a few: Joe Abdallah, Chloe Ahmann, Lindsay Bell, Eeva Berglund, Eric Budde, Craig Campbell, Adrian Currie, Britt Dahlberg, Bahram Dastmalchi, Brian Escobar, Alexander Gordon, Kelly Grotke, Karin Gustafsson, Hugh Gusterson, Doug Holmes, Lara Houston, Zach Howlett, Lei Huang, Janne Hukkinen, Nina Janasik, Timo Kaartinen, Timo Kallinen, Marianna Keisalo, Matti Kojo, Lindsay Krall, Amy Levine, Elina Lehtonen, Markku Lehtonen, Brad Lipitz, Tapio Litmanen, Yash Lodha, Austin Lord, Tim McLellan, Mary Mitchell, Martha Mundy, Brian Pekkerman, Karen Pinkus, Alain Pottage, Beth Reddy, Johan Munck af Rosenschöld, Josh Reno, Minna Ruckenstein, Antti Silvast, Alana Staiti, Magda Stawkowski, Steph Steinhardt, Jake Sullivan, Pedro de la Torre, Juha Tuunainen, John Wagner, Jeff Wall, Anna Weichselbraun, Erick White, Jake Wojtowicz, Rich Zamore, and Malte Ziewitz. Whether they realize it or not, they have each influenced this book in their own unique ways. Any errors are my own.

This book would not have been possible without the sixty-nine fieldwork informants who kindly offered me their perspectives and time. When recruiting interviewees for my study, I suggested they indicate their preference for anonymity on Cornell's "Institutional Review Board for Research on Human Participants" consent form. The idea was that confidentiality would encourage frank and personal dialogue while also protecting their identities. I extend my deepest thanks to all the informants I cannot mention by name. The same goes for the few I can: Timo Äikäs, Mick Apted, Hannu Hänninen, Jehki Härkönen, Satu Hassi, Ari

Ikonen, Tiina Jalonen, Denis Janin, Juha Kaija, Martti Kalliala, Anna Kontula, Kaisa-Leena Hutri, Michael Madsen, Lauri Muranen, Markus Olin, Barbara Pastina, Heikki Raiko, Matti Saarnisto, Timothy Schatz, Johan Swahn, and Juhani Vira. I thank Finnish nuclear waste management company Posiva Oy for giving me formal permission to conduct this research in dialogue with their experts. I thank University of Helsinki's Finnish for Foreigners program for the language training. Special thanks are due to Christa Örn for her Finnish translation assistance.

Finally, over the past few months, Allegra Wrocklage has gone above and beyond: reading over chapter drafts, coming up with clever suggestions for revision, putting up with my daily absorption in writing, and overall being a uniquely smart, fun, and loving partner. I also thank my parents, Vin and Sharon Ialenti, for their love and support, even and especially when my decisions have led me down atypical paths. I dedicate this book to them.

INTRODUCTION
Embracing Deep Time Learning

The word has been out for decades: we were born on a damaged planet careening toward environmental collapse. Yet our intellects are poorly equipped to grasp the scale of the Earth's ecological death spiral. We strain to picture how, in just a few decades, climate change may displace entire populations. We struggle to envision the fate of plastic waste that will outlast us by centuries. We fail to imagine our descendants inhabiting an exhausted Earth worn out from resource extraction and devoid of biodiversity. We lack frames of reference in our everyday lives for thinking about nuclear waste's multimillennial timescales of radioactive hazard. These environmentalist indictments are well rehearsed. Their truths have been spoken to power. Yet our societies, institutions, and intuitions still have not been refashioned to meaningfully think hundreds, dozens, or even a few generations ahead. We need escape routes for breaking out of our shortsighted mental strictures, yet here we are:

Timespans ranging from a few months to a few years determine most formal planning and decision-making—by corporations, governments, non-governmental organisations and international bodies. Quarterly reporting by companies; electoral cycles of 18 months to seven years; planning horizons of one to five years: these are the usual temporal boundaries of our hot, crowded, and flattened little world. In the 1980s, this myopic vision found a name: short-termism.[1]

Some say we have entered a new epoch called the Anthropocene: a new phase in the Earth's history ushered in by our species' own transformations of our planet's climate, erosion patterns, extinctions, atmosphere, and rock record. If this is true, then humans are now a telluric force of nature—agents of not just ecological but also geological change.[2] In this setting, the question is how to become more open to deeper timescales;[3] the challenge is to learn to inhabit a longer now.[4] But intellectually, we are stuck—gridlocked by failures of futurological imagination:

Today, we know our Sun is one of one hundred billion stars in our galaxy, which is itself one of at least one hundred billion other galaxies. But despite these hugely stretched conceptual horizons—and despite our enhanced understanding of the natural world, and control over it—the timescale on which we can sensibly plan, or make confident forecasts, has become shorter. ... There's an institutional failure to plan for the long-term and to plan globally.[5]

This book, I hope, is one tiny step up the mountain before us. Its contention is simple. Gazing into deep time is no longer just for geologists, theologians, paleontologists, astrophysicists, archaeologists, climate scientists, or evolutionary biologists. It is our collective responsibility. Now more than ever, we need the courage to accept our responsibilities as our planet's and descendants' caretakers—millennium in and millennium out—without cowering before the magnitude of our challenge. We need new ways of living, thinking, and understanding our relationships with the Earth's radical long-term. We cannot deny the gloomy reality that from "the perspective of millions of years, the duration of our lived experience, of 'our time,' appears utterly inconsequential."[6] We cannot feel paralyzed by the "seeming insignificance of human purposes within the immense time-span of the universe."[7] If humans are agents of geological change, then thinking in geological timescales is now a pressing ethical imperative. We must learn to learn more and more about the Earth's distant futures and pasts.

Some have already embraced deep time learning. San Francisco's Long Now Foundation is developing a mechanical clock to keep time for ten thousand years to help reframe the way people think and embody deep time. The Finnish Society of Bioart's "Deep Time of Life and Art" project has organized public trips to ancient paintings found on 550-million-year-old rocks located atop Finnish Lapland's two-billion-year-old bedrock.

New Zealand's "A Walk through Deep Time" project has led people along a 457-meter farm fence, meant to represent 4.57 billion years of the Earth's history. They invited geologists, astronomers, biologists, physicists, Maori *matauranga* practitioners, philosophers, and artists to encourage reflection on the Earth's deeper history. Academic institutes like Oxford's Future of Humanity Institute and Cambridge's Center for the Study of Existential Risk have grappled with human extinction scenarios. My former academic home base, Cornell University, has offered an archaeology course that challenged students to imagine the campus as it might look one thousand years from now. The US National Academy of Science has hosted an exhibit called "Imagining Deep Time." Timely books on deep time have recently been published: physicist and science fiction author Gregory Benford's *Deep Time: How Humanity Communicates across the Millennia*, historian Martin Rudwick's *Scenes from Deep Time*, historian Daniel Lord Smail's *On Deep History and the Brain*, paleontologist Stephen Jay Gould's *Time's Arrow, Time's Cycle*, and nature writer Robert Macfarlane's *Underland: A Deep Time Journey*, to name a few.[8]

Deep Time Reckoning approaches these problems a bit differently. I am a cultural anthropologist. I believe the best way to study human relationships is to venture out into the field, meet and interact with people, and try to understand their worlds on their terms. This holds for intergenerational relationships between societies across the millennia, as well. Like many anthropologists before me, I am curious about how different communities grapple with different spans of time, and its past and future horizons. Back in the 1930s, anthropologist E. E. Evans-Pritchard called this "time-reckoning."[9] Studying how people approach time's intervals, scales, and durations can reveal how people understand themselves.[10] That is why, from 2012 to 2014, I conducted ethnographic fieldwork among Finland's nuclear waste disposal experts. This community grappled with deep time on a regular basis.

Nuclear experts routinely deal with long-lived radioactive isotopes. Uranium-235 has a half-life of over seven hundred million years. Plutonium-239 has a half-life of 24,100 years. Doing fieldwork among Finland's nuclear waste professionals, I discovered a world in which mind-bending visions of far future bodies, societies, and environments figure into everyday office routines, policymaking considerations, government and industry plans, and regulatory rules. Along the way, I struggled to come to terms with a seemingly absurd proposition: that these experts could—over the span of several decades—develop increasingly credible forecasts of far future glaciations, climate changes, earthquakes, animal populations, floods, landscape shifts, and more.

When I arrived in Finland, the Finnish nuclear waste management company Posiva was developing what, in the 2020s, will likely become the world's first operating underground disposal repository for high-level nuclear waste. This facility was being built deep in Western Finland's granite bedrock, below the island of Olkiluoto in the Gulf of Bothnia. The Gulf of Bothnia is located in the northernmost arm of the Baltic Sea. Posiva's underground vault was slated to be the final resting place for burying Finland's two commercial nuclear power plants' used-up—or "spent," to use the industry jargon—nuclear fuel. Posiva was also, at the time, in the process of submitting a repository safety assessment report, called a "Safety Case," to the country's nuclear regulatory authority Säteilyturvakeskus (STUK). STUK is the Finnish government's nuclear energy sector overseer and technical rule maker.

The Safety Case's purpose was to convince STUK that the Olkiluoto repository would not pose significant environmental or health risks to future populations. To demonstrate this, Posiva's Safety Case experts worked to forecast the interacting geological, hydrological, and ecological conditions that might surround the repository over the coming tens of thousands, hundreds of thousands, or even millions of years. Intrigued by their Promethean task, I made it my mission to explore how Safety Case experts hashed out relationships between the living societies of the present and the unborn societies imagined to dwell in very distant futures.

For me, the Safety Case became more than just a portfolio of technical evidence supporting Posiva's license application for government permission to construct a geologic repository. It became a window into far future Finlands. I began to realize that the unique tome was chock-full of ideas, strategies, and inspirations for refining how we can better relate to our planet's yesterdays and tomorrows. As denizens of an Earth wracked by ecological crisis, this is our most crucial task. The Safety Case made me wonder: What sort of scientific ethos are Finland's nuclear waste experts adopting in their daily reckonings with seemingly unimaginable spans of time? Have Safety Case experts' efforts to simulate and model distant future ecosystems and geosystems affected how they understand the world and humanity's place within it? What lessons can they offer us in a historic moment some call the Anthropocene? Can learning about their findings help us grow into more skilled deep time reckoners ourselves? If so, how? If not, why not?

To answer these questions, I did not have to take everything the Safety Case experts told me at face value. I did not have to agree with all their opinions, advocate for or against nuclear power, or even believe all their claims about far future Finlands. But I did have to take a small leap of faith. I had to extend some guarded trust that their years of work had left them with a more sophisticated understanding of possible future worlds than I could find in people who had not done the same. I had to listen to the experts carefully, giving them a little leeway and space for explaining themselves. I had to resist any urge to roll my eyes at their multimillennial plans, without first giving them the time of day. Undertaking this anthropological thought experiment was not always easy. I had to learn how to place a measured, self-reflective, cautious faith in the Safety Case experts' deep time knowledge.

But it paid off. I returned from Finland confident that Safety Case expertise can provide guidance for those who seek to transcend rampant shortsightedness. Life today, for many of us, feels frenzied. Computers and tech gadgets require updates at least every few years, if not months. Smartphone fixation distracts minds from long-term contemplation. Throwaway consumerism favors the production of many cheap disposable goods over fewer moderately priced durable goods. Hedge fund managers use computer programs that make financial decisions in seconds or less. They work in digitized global marketplaces and stock exchanges that move even faster. Flexible employment structures, the "gig economy" of contract-based temporary labor, the rise of automation, and "sharing economy" platforms like Uber leave millions anxious about near-term job insecurity. They worry about uncertain future earnings, savings, and benefits. The boom-bust-or-buyout tempos of startup companies are increasingly rapid. Fashion, music, television, and movie trends remain notoriously fickle. And all of this applies only to those with relatively decent standards of living. Billions across the world are confined to even more precarious short-terms: living without ample paychecks or, in more dire conditions, fighting day-to-day to secure shelter, safety, and their next meals.[11]

The Safety Case experts' long-termism runs counter to all of this. In this book, I argue that their techniques must be broadcast far beyond nuclear expert communities. I suggest that this can help us resist shallow time discipline: the powerful cultural, economic, political, and techno-logical forces that interact to fix our attentions on the radical short-term. Today's grandest challenges can be tackled only by taking a longer view. A 2017 United Nations report predicted that the Earth's population will reach 8.5 billion people by 2030, 9.7 billion people by 2050, and 11.2 billion people by 2100.[12] How can these future people be fed, clothed, and housed across decades, centuries, and millennia without vastly ramping up ecological destruction? Species now die out at rates many ecologists say constitute a sixth major planetary extinction event. How can we prevent this irreversible species loss and ensure sustainable life far into the future? A 2018 United Nations report forecasted that 68 percent of people will live in urban areas by 2050.[13] Can architects, engineers, urban planners, computer scientists, and infrastructure builders design smarter cities to adapt to this influx? How might human-induced climate change throw a wrench in their plans? Will tomorrow's supermassive cities become dusty archaeological sites for societies five thousand years from now? Or will our species go extinct long before then?

To begin answering these Anthropocene questions, we must first open our ears to the world's most long-sighted experts. This includes rather obscure experts, like Finland's Safety Case community. That may, at first blush, sound obvious. But such openness to expertise is at odds with another dire global crisis welling up alongside the Anthropocene. For the purposes of this book, I'll call it the deflation of expertise. Today, we are witnessing a rising global skepticism of technocratic knowledge, liberal arts education, scientific research on the environment, and even the very possibility of there being verifiable facts, truth, or a single shared reality out there. In many countries, "experts are increasingly skeptical about publics" and vice versa, as the "bargain" long made between them is "rapidly unraveling."[14] New nationalist and populist movements, on both the political left and right, increasingly challenge the basic notion that "expertise confers legitimacy."[15]

Scientific research may be more sophisticated than ever before. But too often it fails to capture hearts, minds, and imaginations. In this setting, I ask: how could Finland's Safety Case experts dare to offer such ambitious visions of the Earth's far future at a time when populists energize publics around the globe by flagrantly mocking the sober, restrained, often drab voice of the technocrat? How could their confidence in scientific inquiry remain insulated from today's widespread rejections of the mild mannered, meticulous, sometimes pretentious voice of the researcher? How could they persist in their efforts to reckon deep time—unperturbed by the chirping Twitter pundits and lay bloggers who so often drown out the cautious voice of the expert? Is there something unique about the Safety Case experts or Finland's approach to expertise more generally? How could the Safety Case experts plod forward so steadily—undeterred by this growing, worrying, global intellectual crisis?

During the Anthropocene, the silence of complacent experts can be as dangerous as the speeches of any political demagogue. During the deflation of expertise, experts cannot remain fearful of the world outside or cloistered in hidebound ivory towers. But today, many experts who do try to rock the boat—engaging publicly with issues of mass survival—simply fail to make waves. Long-sighted scientists need more platform, visibility, and influence. At the same time, our long-termist learning cannot only be about collecting and synthesizing data from the natural sciences. We need anthropology's spirit of holistic thinking to help us break free from our narrow information silos and echo-chambers. Anthropology can expand our worldviews by helping us figure out where people with whom we disagree are coming from, why they believe what they believe, and how we can learn from their learning. It can remind us that "regardless of what and who we are, we, as individuals and as a society, can dwell in the world in a completely different way from the way we dwell in it at any given moment."[16]

So, I write this book as an exposition of sorts. Chronicling ways that my fieldwork transformed my worldview, it demonstrates Safety Case expertise's power to help us reckon deeper futures and pasts. Pushing back against the deflation of expertise, it advocates a spirit of adventurous learning: of actively seeking out scientifically informed knowledge while embracing an anthropologically informed openness to careful, skeptical, critical listening. This book, then, should not be read as an academic treatise for scholars. It is, rather, a practical toolkit for educated publics, expert and lay alike. I have named the tools in this kit "reckonings." Five or six can be found at the ends of each chapter. They offer open-ended takeaway guidance that I leave behind each step of the way, as I follow Safety Case experts into far future Finlands. Feel free to pick them up and build on them yourself. They are not, after all, difficult instructions telling anyone how to think or act. They are suggested starting points for orienting others in embarking on learning-journeys of their own.

This book's core goal is to encourage as many people as possible to do two things: (1) to pursue independent, expert-inspired, long-termist learning themselves, and (2) to support the highly trained, too-often-ignored, long-termist experts already in our midst. Chapters 1 and 2 focus on the first mission. They walk through mental exercises, thought experiments, and intellectual workouts anyone can do to stretch one's thinking more widely across time. I have derived these brainstorming pathways from my fieldwork among Finland's Safety Case experts. The premise is that, as intellectual and ecological crises take hold, all long-termist knowledge is potentially valuable. We all have something to teach and something to learn. This does not mean that all arguments are equally valid. All information is, of course, not equally accurate. However, digesting many different perspectives—including seemingly irrational, repugnant, or misinformed perspectives—can help us attain more mature, textured, sophisticated worldviews. This spirit of curiosity can help experts and lay-people alike embrace long-termist learning and navigate today's uncertain tomorrows.

Chapters 3 and 4 focus on our second mission. They seek to bolster the special position that long-termist expertise deserves in society today. The question is how to better support long-sighted experts in cultivating, preserving, and disseminating their deep time knowledge. Inspired by the Safety Case community, the chapters' reckonings ask: How can societies empower highly trained, long-sighted experts as their future-gazing guides? How can today's shortsighted private companies, government agencies, NGOs, and universities more fully embrace long-termist learning? Can organizations that employ deep time reckoning experts adopt new policies, programs, and workplace norms to better put their talents into the service of preventing Anthropocene collapse? Can these initiatives help us work against the deflation of expertise by making long-termist expertise more salient among skeptical publics, media pundits, and politicians? Could this help bring deep time thinking closer to the center of societal decision making?

So, that is *Deep Time Reckoning* in a nutshell. But before we set out and start reckoning, it is important that we first take some time to dig deeper into the broader context behind the Safety Case experts' efforts. We'll spend the rest of this introductory chapter collecting background details about the Safety Case community's lifeways, plus the far future Finlands they reckoned. Doing so will give us a clearer view of our case study's bigger picture. Once that is all laid out, I will return to describing the dual crises of the Anthropocene and the deflation of expertise in greater detail. This will help us take a step back and reflect on this book's general approach to expertise, technology, and scientific knowledge. At that point, we will be properly outfitted to embark upon our trek into Finland's tomorrows. But, first, some background to the future.

SEEING DEEP TIME THROUGH OTHERS' EYES

For thirty-two months from 2012 to 2014, I lived in Finland. A parliamentary republic of around 5.5 million people, Finland is the most sparsely populated country in the European Union. By area, it is slightly smaller than Germany, slightly larger than Poland. Sweden is to its west, Russia to its east, Estonia to its south, Norway to its north. Finland borders the Gulf of Bothnia, the Gulf of Finland, and the Baltic Sea. It is a highly industrialized, mostly free-market economy with a high per capita GDP and an extensive Nordic welfare state. About 90 percent of its population speak Finnish as their first language. Just over 5 percent are part of Finland's Swedish-speaking linguistic minority. Finland is also home to Saami, Roma, Estonian, and Russian minorities, plus refugee populations from Somalia, Iraq, Afghanistan, and elsewhere. It is known for its telecommunications, electronics, metals, wood, engineering, electronics, clean tech, and information technology industries. About 30 percent of Finland's energy production comes from its domestic nuclear power plants.

To conduct fieldwork, I took Finnish language courses, immersed myself in Northern European life, conversed with Safety Case experts in person, read their technical reports, and heard out their critics. I visited them at work and grabbed coffee, lunch, or drinks with them whenever possible. We discussed life in Finland, what captured their imaginations at work, how they organized their time, and how their time was organized for them.[17] I picked their brains about how they grappled with futures near and deep. I chatted with physicists, engineers, geologists, mathematicians, hydrologists, artists, computer modelers, industry lobbyists, managers, chemists, finance professionals, activists, lawyers, politicians, academics, and others. I ended up recording 121 interviews with people who worked on or had something special to say about Finland's nuclear energy and waste initiatives.

My fieldwork informants worked at Finland's nuclear regulatory authority STUK, the nuclear waste company Posiva, the Technical Research Center of Finland (VTT), the Geological Survey of Finland (GTK), engineering consultancy Saanio & Riekkola Oy, lobbying association Confederation for Finnish Industries, clay technologies consulting firm B+Tech, Finland's Parliament (Eduskunta), Aalto University, Greenpeace, lobbying association Finnish Energy Industries, Friends of the Earth, power companies Fennovoima, TVO, and Fortum, the general public, and elsewhere.[18] Some worked on the Finnish Research Programme on Nuclear Waste Management, which oversaw the country's nuclear waste knowledge base. In this book, I mask my informants' identities with pseudonyms. Preserving research subjects' anonymity is a common practice among anthropologists working with sensitive or personal situations.

When I moved to Finland, Posiva had already dug a four-kilometer access tunnel deep under Olkiluoto, which led down into a subsurface lab called Onkalo (Finnish for "cavity" or "hollow"). Onkalo researchers provided Safety Case experts with information about Olkiluoto's underground conditions. Onkalo was to be renovated into a repository and expanded to a depth of about 400 to 450 meters. At the time, the plan was to make space for storing up to 9,000 tons of radioactive waste—though, more recently, the facility's initial capacity was lowered to 6,500 tons. Posiva planned to bury Finland's spent fuel "bundles" there. By late 2015, Finland's nuclear reactors had left behind around 13,500 of them.

The bundles contained over 2,000 tons of uranium, plus fission products like cesium, iodine, and technetium.[19] Posiva's plan was to slide the bundles into tube-shaped cast iron "insert" containers. They would then seal the inserts inside large copper canisters at a nearby encapsulation plant. Once placed in one of the repository's underground holes, the canisters would be surrounded by an absorbent bentonite clay. The clay would serve as a "buffer" between the canisters and bedrock. This repository approach, known as KBS-3, was mostly derived from Swedish nuclear waste company SKB's designs. Still, Posiva's repository was expected to open for business first, sometime in the 2020s. After almost one hundred years of burial operations, the tunnel would be backfilled, sealed off, and decommissioned around 2120.

In the meantime, Finland stored its spent nuclear fuel bundles in temporary cooling pools near the two power plants that generated them. One of the plants is owned by the Finnish energy company Fortum, located in the Loviisa municipality on Finland's southern coast. Loviisa's two reactors were turned on in 1977 and 1980, respectively. Fortum owns 40 percent of Posiva. Another plant is owned by the Finnish nuclear power company Teollisuuden Voima Oyj (TVO), located next to Posiva's repository site on the island of Olkiluoto in Western Finland's Eurajoki municipality. TVO's two reactors were turned on in 1979 and 1982, respectively. TVO owns 60 percent of Posiva. TVO's and Fortum's spent fuel needs to cool in the pools for about forty years for its radioactivity and heat to fall to levels suitable for encapsulation and burial. Since 1983, Finland's national policies have required companies that generate spent nuclear fuel to take responsibility for managing and disposing it. When I left the country in late 2014, Finland's State Nuclear Waste Management Fund had accumulated over 2.3 billion euros from the annual fees that TVO and Fortum had paid into it over the years. This fund, mandated by Finland's 1987 Nuclear Energy Act and overseen by its Ministry of Economic Affairs and Employment, supports the work of Posiva.

Within Finland's nuclear waste management community, the Safety Case experts' mission was straightforward. They were to develop evidence-driven methods for convincing Finland's famously strict nuclear regulator STUK that any exposures delivered to future populations would be safely below the country's legal limits—including populations living tens and hundreds of thousands of years from now. Once STUK approved Posiva's Safety Case, Finland's Ministry of Economic Affairs and Employment would issue Posiva a construction license for the facility. The Safety Case experts worked to forecast whether and how radionuclides could someday escape from the repository, travel through underground rock fractures and groundwater channels, get released at various points on Western Finland's surface, and then disperse in aboveground ecosystems—bioaccumulating, at times, in plants and animals along the way. Their big question was whether a release could expose human or nonhuman lifeforms in the Olkiluoto region to radiological risks, in unlikely far future scenarios.

To develop their portfolio, some Safety Case experts prepared reports with titles like *Climate Scenarios for Olkiluoto on a Time-Scale of 120,000 Years*. Others examined how the coming Ice Age's potentially three-kilometer thick ice cover could affect the facility 50,000 or 60,000 years from now. They envisioned postglacial seismic events that could occur as the ice sheet melts and retreats. Some researched whether groundwaters of various chemical compositions could corrode the repository's KBS-3 "engineered barrier system." Others simulated Western Finland's lakes, rivers, mires, and forests sprouting up, disappearing, and changing shape and size over the next 10,000 years. They considered future soil erosion, floods, and fires. Some experts made computer models of how the Olkiluoto *island* site will eventually become an *inland* site. They modeled how Finland's shoreline will widen further into sea, continuing to rebound upward in elevation. It has been doing so since the retreat of the previous Ice Age's glacial ice sheet. This process of land "uplift" will eventually engulf Olkiluoto into Finland's mainland.

Other Safety Case experts investigated questions such as: What complications could the decaying radioactive waste's heat pose to the repository's long-term performance? At what rate will Posiva's copper canisters and cast iron inserts corrode underground? What will the effects of far future permafrost be? How will the coming Ice Age's glacial ice cover affect the repository? What about the potential seismic activity slated to occur once the ice sheet melts? What role will anthropogenic climate change play in all of this?

Appreciating the ordinary details of the Safety Case experts' lives was key to establishing relationships of trust between anthropologist and informant. This meant considering how some were grandparents, others divorcees, others aloof loners, others rambunctious partygoers, while still others had recently become parents. One physicist in his fifties was training for an iron man endurance competition. Some spent summers crunching numbers in their office. Most enjoyed time in the Finnish countryside at their family's *kesämökki* ("summer cottage"). One's daughter had a pet hedgehog named Nipsu. Another decorated the wall near his desk with images of his Finnish Lapphund he printed off from his computer. One read Wittgenstein's *Tractatus Logico-Philosophicus*. Another read works by Bruno Latour, Mary Douglas, Ulrich Beck, Alice Munro, and Marcel Proust. A security guard at a building housing a research nuclear reactor was a fan of the late-1980s early 1990s action-adventure television series *MacGyver*. In their childhoods, two experts, both geologists today, dreamed of someday becoming archaeologists. A high-status regulatory expert grew up fascinated by what she saw as the more ecologically attuned lifeways of Native Americans. A geologist went boxing with her daughter on Tuesdays to keep her arms, back, and shoulders moving after hunching over her computer all day. It took a village to make far future Finlands appear.

With our rapport established, Safety Case experts began showing me how they wove together powerful visions of far future Ice Ages, seismic events, erosion processes, human and animal populations, and climate changes. Our open-ended conversations were often speculative and philosophical in spirit. They taught me how to see deep time through their eyes, and drew me to reimagine how the Earth might look a decade from now, a century from now, millennia from now, and even millions of years from now. They taught me how to reckon deep time not as a matter of sci-fi futurism or navel-gazing rumination, but as a practical labor of regulatory science, government policy, and corporate planning.

Moving forward with my thought experiment, I began to ask: Could it be that nuclear waste disposal experts—long approached with skepticism by environmentalists and critical academics like me—are developing among the best resources for stretching our intellects into ever more distant futures? Could repository safety assessments inspire us to extend our intellectual horizons further forward and backward across time? What if their forecasts were translated from drab calculations and technocratic jargon into more lively and accessible prose? Could opening ourselves to deep time in ways inspired by Finland's Safety Case experts motivate positive change in our ways of living on a damaged planet? Could societies opening themselves to expert-inspired, long-termist learning help us counteract the deflation of expertise?

To begin answering these questions, we must first think in terms of a bigger picture about Safety Case expertise's broader implications. So, with Finland's nuclear waste repository project in mind, let's return to the Anthropocene. This is, after all, the key historical setting that makes embracing deep time learning all the more essential.

ENTER THE ANTHROPOCENE

The Anthropocene is, as I write, not (yet) a formal interval of geological history. But the idea behind it is not new.[20] The view that humanity has affected the Earth so profoundly that it has become a formidable geological force has gained considerable ground. The atmospheric chemist and Nobel Prize–winner Paul Crutzen revived the Anthropocene concept in 2000, advocating for it at an International Geosphere-Biosphere Programme conference in Mexico.

In 2012, the International Union of Geological Sciences (IUGS) established an Anthropocene Working Group to vet the proposal, and in 2019, the taskforce voted to recognize it as a formal geological epoch that originated around 1950. However, it must surmount several more hurdles before the Anthropocene could ever be recognized as a formal epoch on the geological timescale: the official scientific chart for dividing our planet's deep history into eons, eras, periods, epochs, and ages. It first requires a supermajority of votes from both IUGS' International Quaternary Subcommission and the International Commission on Stratigraphy. It also requires IUGS' executive committee's approval. It is often said that evidence must be found, in stone and sediment, indicating that far future stratigraphers will be able to look at the rock record below their feet and in the ocean beds to measure a clear break from the previous epoch. But paleontologist Jan Zalasiewicz, the chair of the Anthropocene Working Group, has said the question was less about detecting human influence in rock strata; for him, it was about revealing a fundamental, human-induced change in the Earth system.[21]

The Anthropocene idea, however unofficial, has already sparked lively scientific debates. Like Safety Case research, it can stretch our minds across time. One question is when it should be decided to have begun. Crutzen originally suggested the Anthropocene began in 1784 with the invention of the steam engine, which set into motion centuries of widespread fossil fuel addiction. In 2011, however, he and his colleagues proposed 1800 as a more logical start date. That is about when the Industrial Revolution became entrenched in Great Britain and elsewhere.[22] Other start date proposals look to the Great Acceleration: the post–World War II explosion of automobiles, synthetic nitrate fertilizers, airplanes, population booms, international tourism, carbon emissions, nuclear arms race, leaded and concrete petroleum, plastics, global telecommunications, national GDPs, urbanization, energy use, water use, paper production, and so on. This data spike coincided neatly with another proposed start date: the 1945 advent of nuclear weapons testing, which dispersed new human-made radionuclides across the planet. The Anthropocene Working Group has seen the Atomic Age as offering far future stratigraphers a clear signature of materials that did not exist prior to humanity taking the helm as a geological force.[23] Critics, however, worry that this nuclear origin story would let the Anthropocene's primary driver, fossil fuels, off the hook too easily. Meanwhile, alternative start dates have been proposed: the mass extinctions of Pleistocene megafauna, the Neolithic Revolution's advent of agriculture, and the European conquest of the Americas, to name a few.

The Anthropocene idea has sparked long-sighted debates among social scientists and humanities scholars. Historian Dipesh Chakrabarty has argued that the Anthropocene dissolves the long-standing divide between the "natural" history of our planet and the "human" history of peoples, merging them into a single "geo-history."[24] He has noted that scholarship on postcolonialism, theories of globalization, and Marxist critiques of capitalism scarcely prepared him to grasp this geo-history's significance. Literature scholar Timothy Morton has cast the Anthropocene as emerging from the "logistics of agriculture" that arose in the Fertile Crescent ten millennia ago.[25] Anthropologist and philosopher Bruno

Latour has argued that the Anthropocene idea bypasses the modern divide between nature as a stable, slow, external setting in which culture—our human dramas and lifeways—is imagined to take place. As a result, in the Anthropocene, when one encounters a "natural" phenomenon like a waterfall or soil, one also encounters *anthropos*—the human—and vice versa. This postnature condition, Latour argues, makes politics and science "assume a totally different shape" that calls for them to no longer be seen as opposed to one another.[26] This blurs boundaries between the social and natural sciences.[27] Sociologist Nikolas Rose has called for a "critical friendship" between the social and life sciences.[28] In the Anthropocene, collaborations between anthropologists and long-sighted hard scientists like Finland's Safety Case experts are essential.

So, the *Economist*'s 2011 proclamation "Welcome to the Anthropocene" was certainly premature. But the idea has proven worthy of serious attention. In this book, I take it as an impetus for population-wide, expert-driven, long-termist learning. This is a crucial mission even if the Anthropocene is never inaugurated as an official epoch. The multiple, enormous, overlapping environmental crises that the idea has pulled together under a single umbrella are reason enough for embracing a sea change in how we reckon with futures and pasts. However, if humans *are* judged to be agents of geological change, then the value of better syncing our everyday thought patterns with deeper timescales may become clear to thousands, millions, or even billions more.

To achieve all this, though, we must face a sobering reality. Thinking about deep time is extraordinarily new for us; our task will not be easy. Even the most highly educated scientists are, historically speaking, rookies at it. For Europeans in the Middle Ages, the whole cosmos of human history covered only a few short millennia, from the Earth's creation to its prophesied apocalyptic end.[29] Until the nineteenth century, most Westerners assumed the world began only a few thousand years ago. The year 4004 BCE was widely seen as the biblical dawn of time, as defined by Irish Archbishop James Ussher.[30]

Since then, Sigmund Freud has described what he saw as modern science's three most humbling revolutions for humanity. This first was Copernicus's sixteenth-century revelation that our planet is not the center of the universe (which led scientists to realize that the Earth is a minuscule speck among countless other celestial bodies). The second was Darwin's nineteenth-century theory of evolution (which placed the human species within, not above, the animal kingdom). The third was Freud's own nineteenth-century discovery of unconscious psychological drives (which showed humans to be far less rational than previously assumed). In 1987, paleontologist Stephen Jay Gould suggested we should add a fourth humbling revolution to the list: the rise of the geological sciences. The geologic timescale relegated the whole human story to a tiny moment in the deeper history of our planet. This revolution in time reckoning is often attributed to James Hutton's 1780s book *Theory of the Earth* or Charles Lyell's 1830s book *Principles of Geology*. The jury is still out on whether human societies have processed these profound insights—whether we have truly come to grips with our profound insignificance.

Yet, even as we take this hard look in the mirror, we must not approach the Anthropocene idea uncritically. Social scientists and humanities scholars have criticized Anthropocene advocates for implying that a single unified humanity is to blame for centuries of environmental devastation. After all, what does one really mean by saying that "humanity" is now a geological force, given that paleohistory, microbiology, biochemistry, and other fields define "human" so differently?[31] Can *Homo sapiens* as a whole really be blamed when one US citizen emits more carbon into the atmosphere than five hundred citizens of Ethiopia or Burundi? If 19 million New York State residents consume more energy than 900 million from sub-Saharan Africa, why lump them all into the same basket?[32] Why not call out capitalism as a culprit by renaming the Anthropocene the Capitalocene?[33] Or call out the one-percenters by renaming it the Oliganthropocene?[34] Given that mostly white, mostly male Western humans originally adopted the steam engine, why not call it the Chthulucene—referencing H. P. Lovecraft's "misogynist racial-nightmare monster"?[35] Finally, does presenting the Anthropocene as a new revelation let past polluters off the hook too easily? They have, after all, systematically ignored environmental warnings from the nineteenth century and before.[36]

These critiques are valid. However, my goal is not to hype up the world-historical significance of the Anthropocene. Nor is it to pull the rug out from under it by pointing out the very real baggage that comes with it—as important as tasks these are. My aim is simply to stress that, regardless of who got us into this ecological mess, the Anthropocene idea can inspire, agitate, or compel virtually anyone to cultivate their deep time thinking skills. This is a good thing. So, when I use the pronoun "we" in this book, I refer not to humanity's undifferentiated *everyone*. My "we" refers, simply, to *anyone* who finds oneself seeking a longer-term perspective at a time when so many social scientists, journalists, natural scientists, humanities scholars, museum curators, and others find fascination in deep time. This "we" refers to anyone who wishes to better sync their thinking with deep, geological, Anthropocene timescales. It refers to anyone who wishes to resist the deflation of expertise by promoting greater societal openness to long-termist learning.

This book uses the word "reckoning" in a few different ways, too. It draws from its multiple dictionary definitions. The first refers to what the Safety Case experts did: how they reckoned deep time by calculating, estimating, and drawing conclusions about far future Finlands. The second refers to what I do: I reckon with the Anthropocene planetary crisis and the deflation of expertise intellectual crisis by concluding each chapter with reckonings: "opinions or judgments," "ways of thinking," and "considerations" about how we can better foster long-termism. The third definition refers to the general spirit of my anthropological thought experiment: a "settlement," or even indictment, of "past mistakes or misdeeds" done to our planet by people that came before us. This book is, in this sense, a reckoning with legacies of the past.

With these reckonings in mind, let's dig a little deeper into the Safety Case experts' professional culture. We now have a richer sense of the Anthropocene as a backdrop to this case study. So, let's meet a Finnish nuclear waste expert and learn more about how and why he grapples with deep timescales. This will help us better understand his expert community's approach to creating futurological knowledge. It will also help us appreciate the nuances of Safety Case experts' scientific sensibilities before we return to exploring the deflation of expertise in greater detail.

SISYPHEAN ANTS

It was summertime in Helsinki in 2013. I was sipping coffee in an office with Risto, a computer modeling expert involved with Posiva's Safety

Case projects. When Risto spoke, he strung his sentences together carefully and methodically. His tone was calm, warm, and straightforward. He wore a simple, casual, dark-colored collared shirt to work. He recounted a childhood memory to me. Risto remembered sitting outside, decades ago. He was watching an ant repeatedly trying to climb up a ridge of mud formed by a human bootprint in the ground. The ant kept climbing up, falling down, trying to climb up again, falling down, *ad nauseam.* Risto was never sure whether the ant ultimately made it. Yet he was struck by two things. The first was the Sisyphean ant's tenacity in ceaselessly trying to surmount what should have been, perhaps like the Safety Case's efforts to augur far futures, dismissed outright as an insurmountable obstacle. The second was the steadfastness with which the ant repeatedly climbed up and fell down: deploying its best problem-solving strategies to tackle an inordinately difficult challenge without any certainty it could ever be accomplished.

Some of Risto's colleagues identified with ants too. They expressed feeling like simple ants working in a larger Safety Case ant colony. They felt themselves inhabiting a collective intelligence that superseded any individual expert's intelligence. Personal reflections like these helped me see what the Safety Case's deep time reckoning was all about. It was about pressing onward toward impossible scientific horizons while working in complex collaborations that, as a whole, exceed any single person's comprehension, yet still somehow work. We hope.

Risto, like many Safety Case experts, was a consultant with a PhD. He was employed by a state-owned nonprofit research institute called the Technical Research Center of Finland (VTT). He received his share of Finland's State Nuclear Waste Management Fund money through his contracts with Posiva. Some of Risto's colleagues worked directly for Posiva, which employed about ninety people. Posiva was headquartered in Eurajoki municipality near Olkiluoto, but also had an office in Helsinki. Posiva contracted with many experts based in Helsinki and nearby Espoo. Its consultants were employed by VTT, the Geological Survey of Finland, private consultancies such as Pöyry Oyj, Saanio & Riekkola Oy, clay technologies consultancy B+Tech, or companies abroad like Sweden's risk assessment firm Facilia AB.

The Safety Case experts worked within a distinct project management hierarchy. When I began fieldwork in 2012, they were managed by a research director, who was a member of Posiva's eight-person operations team. That team reported directly to Posiva's president. Posiva's management group was overseen by a board of directors, composed of higher-ups from power companies TVO and Fortum. The Safety Case experts were also overseen by the Finnish Research Programme on Nuclear Waste Management (KYT). KYT was an independent organization tasked with understanding Finland's total nuclear waste knowledge base across several organizations.

Risto's job was to make models of how and whether radionuclides would or would not move around in Western Finland's underground, if a copper nuclear waste canister were to rupture someday. His models were part of Posiva's Safety Case portfolio. The Safety Case consisted of thousands of pages of evidence detailing the repository's design, reporting engineering principles, and presenting numerous quantitative models, computer simulations, and scenarios forecasting its fate over the coming millennia.

Risto spent most of his time developing geophysical and hydrological models in a Helsinki office. Other Safety Case experts worked outdoors—drilling boreholes and taking samples of water and minerals. Some studied the bedrock deep in Onkalo. Others mixed waters and clays together in surface-level laboratories in Helsinki. The Safety Case "biosphere assessors" made simulations of 6,000 future people living around the Olkiluoto area. All inhabitants were assumed to eat only local food. They determined that a certain amount (a "critical group") may face a greater likelihood of radionuclide exposure as a result of future unlucky food selections. It was obvious to the biosphere assessors that the region could not support a population of that size without importing food from elsewhere. It is freezing and dark there for much of the year. However, Safety Case experts made "conservative" assumptions like these to emphasize how Olkiluoto's hypothetical future people would be safe, even in worst-case or impossibly pessimistic scenarios. This was done with reference to STUK regulatory rules like this one:

The dose constraint for the most highly exposed individuals, 0.1 mSv per year, stands for the average individual dose in a self-sustaining family or small village community living in the environs of the disposal site, where the highest radiation exposure arises through different exposure pathways. In the living environment of this community, a small lake and a shallow water well, among other things, are assumed to exist. ... The average annual doses to larger groups of people living in the environs of a large lake or sea coast shall also be addressed.[37]

When I chatted with Risto, he was just one man sitting in a chair. But the group collaborations, professional networks, and organizations in which he worked had long molded his thinking. Risto's nuclear knowledge was always partially an "effect" of the "bureaucratic machinery" that helped forge it.[38] The Safety Case had emerged from clusters of domestic laws like Finland's 1957 Atomic Energy Act, 1987 Nuclear Energy Act, 1988 Nuclear Energy Decree, 1991 Radiation Act, and Government Decree 736/2008. It was designed to conform to regulatory requirements in STUK's highly technical YVL (for *Ydinvoimalaitos*, or "nuclear power plant") regulatory guides. It was influenced by Sweden's SR-Site repository safety assessment. It was made with global standards defined by international bodies like the International Atomic Energy Agency and the OECD's Nuclear Energy Agency in mind. It considered technical recommendations from the UN Scientific Committee on the Effects of Atomic Radiation—plus the International Commission on Radiological Protection.

Risto's scientific prowess was, in part, his own. At the same time, it was also the fruit of a vast landscape of international institutions, industry initiatives, government programs, university institutes, and legal entities spread across the globe. From this professional landscape sprouted visions of distant future worlds. Safety Case experts' quantitative projections of far future radionuclide releases into Western Finland's ecosystems focused on the 10,000 years following the repository's scheduled closure in 2120. My informants explained how this was roughly the amount of time since human habitation began in Finland after the previous Ice Age. Other analyses looked hundreds of thousands of years into the future. The "reference period" for modeling how the Olkiluoto repository's physical architecture would evolve was set at 250,000 years, to include at least one ice age cycle. Some Safety Case experts looked much further ahead. They worked on Posiva's scenario, *The Evolution of the Repository System beyond a Million Years in the Future.* Doing so stretched their minds widely across time. Engaging with them anthropologically can help us stretch ours.

But, to get there, we must first learn more about how Risto's deep time visions emerged from the humdrum here-and-now of office life, scientific projects, short-term schedules, and deadlines. We must learn to approach the Safety Case experts not just as specialists with technical knowledge. We must approach them as humans with dreams, hobbies, anxieties, hopes, frustrations, quirks, passions, gossip, regrets, kindnesses, and opinions. We must suspend any preconceptions we have about nuclear waste's deep timescales. We must commit to approaching them, temporarily at least, on Safety Case experts' terms. Only then can we understand just how starkly their ambitions diverged from the spirit of today's deflation of expertise.

WHEN DEEP TIME BECOMES SHALLOW

Before embarking for Finland in December 2011, I watched Danish filmmaker Michael Madsen's 2010 documentary about the Olkiluoto repository, *Into Eternity: A Film for the Future*.[39] The film painted Posiva's efforts to reckon deep time with aesthetics of desolation and bleakness, of austere machinery and industrial processes. Madsen portrayed the project as a place where dark souls tended to the world's most lethal waste in a lifeless cave beneath a frigid island at the edge of the habitable world. The project was depicted as a place of gloom and gravity, stillness and darkness. With this scene set, Madsen told a story of nuclear waste experts speaking straightforwardly of their plans to engineer an underground facility resilient to the contingencies to befall the Olkiluoto region over the coming millennia. The moods, ambiances, and cadence of the film were stirring. The story it told was engrossing. Deep time appeared mysterious, overwhelming, awe-inspiring, and otherworldly.

For Madsen, Finland's nuclear waste invited philosophical speculation about the ethics of pollution, human extinction, communication between societies across millennia, and civilization's grandest aspirations. This had precedents in academia. Ethicist Kristin Shrader-Frechette had already reflected on how nuclear waste's multimillennial half-lives pose intergenerational ethical dilemmas.[40] These half-lives have challenged scientists to model very complex, contingent, and interconnected ecosystems, engineered facilities, and geological systems. Social scientists had also already analyzed the US Yucca Mountain high-level nuclear waste repository project's radically long-term engineering demands.[41] Physicist and science fiction author Gregory Benford had described the United States' challenge of designing monuments to warn far future societies to stay away from the WIPP transuranic nuclear waste repository.[42] WIPP is located deep in a New Mexico salt deposit. These societies, Benford explained, may have wildly different languages and signage systems than we do. In 2015, these questions resurfaced in Harvard professors Peter Galison and Robb Moss's documentary about containing nuclear waste and "imagining society 10,000 years from now."[43] In 2018, historian Gabrielle Hecht summed up these challenges elegantly:

It is not just that ten thousand years (or longer—plutonium's half-life is twenty-four thousand years) exceeds human *design* horizons. That sort of time scale exceeds human *language* horizons. How to make waste repositories legible to humans millennia in the future? How to persuade our descendants that the buried materials are permanent waste, too dangerous to ever be revalued? Such projection into the future does not just require reckoning with geology; it also requires reckoning with language and representation. Attempts to address such questions have involved anthropologists, archaeologists, philosophers, artists, and linguists. Interdisciplinary committees have imagined two- and three-dimensional signage to warn future generations. Inevitably in such discussions, someone invokes the Giza pyramids to demonstrate that taboos against plunder stop working after a while.[44]

In the late 1980s and early 1990s, the US Department of Energy assembled interdisciplinary committees like these to brainstorm warning monument designs for the WIPP repository. Some experts proposed carving a foreboding note into a huge concrete slab in seven different languages, from Navajo to Chinese to English. The goal was to scare off future tomb-raiding treasure hunters, archaeologists, local communities, and miners. Part of the slab would read:

This place is not a place of honor … no highly esteemed deed is commemorated here … nothing valued is here. What is here was dangerous and repulsive to us. This message is a warning about danger. … The danger is still present, in your time, as it was in ours. The danger is to the body, and it can kill. The form of the danger is an emanation of energy. The danger is unleashed only if you substantially disturb this place physically. This place is best shunned and left uninhabited.[45]

The experts also proposed that "ominous and repulsive" non-language-based signs should mark the site, in case the tongues of today die out. One idea was to surround the landscape above WIPP with fields of enormous spikes and frightening thorns. Others suggested warning signs depicting the eerie, agonized face of Edvard Munch's painting *The Scream*. A decade before that, members of the US Department of Energy's and Bechtel Corporation's "Human Interference Task Force" proposed bioengineering living cats that change colors in response to radiation. After releasing the cats into the world, the goal would be to spread around folkloric fairy tales, poems, myths, and paintings. These modes of message transmission have endured through the ages. They would signal to future generations that seeing oddly colored cats means danger is near. Around the same time, semiotician Thomas Sebeok proposed a hypothetical council of experts called the "Atomic Priesthood." The priesthood would preserve knowledge of nuclear waste sites and relay warnings about them through myths and rituals. This plan was inspired by the Catholic church's success in conveying messages for over 2,000 years.

This is all pretty mindboggling. So, perhaps it is unsurprising that Madsen took a lengthy detour in his film about Finland's Olkiluoto repository project to discuss the United States' Yucca Mountain and WIPP repository projects' warning monument ideas. Yet, as I got know Safety Case experts like Risto, I found their imaginations piqued by a rather different set of questions. My informants were caught up in the scientific, regulatory, and engineering details of their work. They focused on technicalities and scientific uncertainties more than any philosophical reverie on deep time's forbidding expanses. Many abstained from speculating about future human signage systems. Their interests diverged from those of many social scientists, journalists, and humanities scholars. My informants never quoted philosophical works on the sublime or the uncanny. They never mentioned deep time's ghostliness.[46] Few cared to discuss how "the mind seem[s] to grow giddy by looking so far into the abyss of time."[47] Nobody, to my knowledge, lost sleep over how viewing our lives from the "meta-level gaze from geological time" can leave us feeling like the "individual, the community, nation, history, creed, institution, and religion" are insignificant.[48] Some explained that, given Posiva's repository R&D work's urgency, they simply could not afford to get lost in the existential dread of Madsen's ruminations. Others critiqued how the sensationalism surrounding repository warning monument proposals had deflected public attentions away from their more substantive work on geoscience, engineering, regulation, financing, and systems modeling.

After months of fieldwork, I began to see how and why the flashy popularity of these issues had distracted cultural analysts like me. They had shifted attention away from the just-as-fascinating but less-well-known far future visions found in seemingly boring regulatory science documents like Finland's Safety Case models. The same went for my home country's equivalent: the US Department of Energy's stagnated Yucca Mountain repository project's License Application's million-year Total System Performance Assessment models.[49] Doing anthropological fieldwork altered how I saw, related to, and wrote about nuclear waste's deep time. The assumptions I brought to Finland with me broke down.

Safety Case experts' professional worlds turned out to be worlds apart from those in Madsen's film. Most of my informants worked in modest-but-comfortable office buildings. Their offices were adorned with fluorescent lights, coffee machines, clean cafeterias, saunas, prints of artwork on the walls, and unostentatious brick exteriors. Many sat in chairs for much of their workdays. They ran models on computers, scrutinized regulatory requirements, and pored over datasets. Sometimes they stayed up late into the night to finish technical reports before a deadline. They spent most of their time working quietly. They chatted lightly and joked among themselves. They attended meetings, drank coffee, and looked over reports. This deromanticized what I had read in books and watched in films about nuclear waste's deep time. As time passed, I found myself focusing more on how Safety Case visions of far future Finlands sprouted from rather short-term timespans: deadlines, schedules, funding disputes, project phases, career stages, daily plans, five-year plans, contingency plans, human life courses, and interpersonal dramas. Nuclear waste's deep time's aesthetics of gloom, awe, and profundity faded from view. They were displaced by the Safety Case project's here and now. Deep time became shallow.

I was surprised to discover that, even as I engaged with very alien far future worlds, most of the "reckonings" I collected ended up pertaining to some of the most common features of human experience. These included the power of analogy (chapter 1), the power of pattern-making (chapter 2), the power of shifting and reshifting perspectives (chapter 3), and the problem of human mortality (chapter 4). But these familiarities, I realized, can be useful. Their sheer relatability can serve as a launching-off point for us to pursue long-termist learning ourselves.

As mundane as the Safety Case experts' office lives were, their optimism about human ingenuity's power to secure better ecological tomorrows never ceased to shine through. Theirs was a more cautious, measured, reserved techno-optimism than that of, say, proposals to fix Anthropocene problems using geoengineering. Geoengineering proposals have included plans to put fertilizers into oceans to raise their carbon dioxide uptake, or to pump reflective particles into the atmosphere to deflect sunlight.[50] Most Safety Case informants were eager to protect the Earth from radiological hazards, but were more skeptical of humanity's environmental impacts overall. They did not celebrate what some "ecomodernist" thinkers call a "great Anthropocene," in which humans will learn to "use our extraordinary powers to shrink our negative impact on nature."[51] Instead, they placed their trust in something simpler. They trusted that a highly disciplined, tightly organized, well-trained, adequately funded group of experts can respond pragmatically to the nuclear waste problem. They trusted that this group could progressively improve the quality of their solutions as the months, years, and decades went on. Public appreciation of this softer, subtler form of techno-optimism can help us escape the worrying grasp of the deflation of expertise.

Yet I was not able to appreciate the Safety Case experts' modest faith in scientific reasoning until my fieldwork ended. When I returned to Cornell University in the United States, I realized that their practical, unassuming, yet wildly ambitious endeavors to reckon far futures were out of step with the grave crises of expertise that had come to afflict so many other twenty-first-century societies. The relatively strong public trust that Finnish publics placed in Posiva's work felt like it came from some strange other world. The mild reverse culture shock of returning home did not conclude my learning-journey. It merely opened up a new chapter. I grew concerned that a brewing global intellectual crisis would exacerbate the Anthropocene's brewing ecological crisis. To give a richer sense of what I mean, let's take a closer look at the deflation of expertise and its many implications.

THE DEFLATION OF EXPERTISE

In August 2014 I was back in upstate New York. I had just returned home from Finland. It was time to turn my fieldwork findings into a doctoral dissertation. I began to see the extent to which the authority, credibility, and confidence of many kinds of experts had been attacked while I was away. For many, faith in financial experts had been damaged by failures to predict the 2007–2008 global financial crisis. For many others, faith in experts who had been proclaiming a coming "nuclear renaissance" since the early 2000s—in which emissions-free nuclear energy was to become a twenty-first-century climate solution—had been sapped by failures to stop Japan's 2011 Fukushima meltdown and by wariness of nuclear reactor technology's high capital costs. For some, faith in national security experts had been shaken by failures to foresee the 9/11 terror attacks. For others, faith in military experts had flagged since the United States invaded Iraq based on false pretenses about weapons of mass destruction. For some, faith in intelligence experts had been stunted since Edward Snowden publicly leaked the US National Security Agency's massive global surveillance operations. For others, faith in big data science was losing its luster, now associated with digital privacy losses and surveillance capitalism.

Skepticism was surely warranted in the wake of these grave failures of expertise and breaches of public trust. However, as the months and years went on, all sorts of claims to expert knowledge were increasingly written off as mere pompous elitism. Come the 2016 presidential election, nationalist-populist movements would clamor loudly against technocrats, bureaucrats, globalists, and the so-called deep state. Crime and jobs statistics, polling methods, vaccine science, government reports, university funding, pandemic disease alerts, and environmental research would be questioned. I found myself thinking that Finland's Safety Case experts' aspirations to reckon deep time would, in these contexts, likely be mocked as naïve or hubristic.

As I write in 2020, crises of expertise are ongoing. Stem cell research, economic models, climate experts, critical social theorists, cell phone radiation, and human evolution are targeted in frenzied social media free-for-alls. A 2012 *Scientific American* article described an "anti-intellectual conformity" gaining ground at both ends of the US political spectrum. This was "precisely the moment that most of the important opportunities for economic growth, and serious threats to the wellbeing of the nation, require a better grasp of scientific issues."[52] By 2017, it was no longer edgy or radical for a postmodern philosopher or critical sociologist to critique truth, reality, or scientific rationality. "Alternative fact" and "fake news" had become mainstream buzzwords. *The Oxford English Dictionary* named "post-truth" its 2016 Word of the Year. In April 2017, science became something to march on Washington for. I felt nostalgic for the Safety Case experts' day-to-day faith in facts and reason. I worried the United States was too distracted by noisy clutter to prep for ecological destruction. Deluges of clickbait web content and twenty-four-hour news cycles left netizens lost in jungles of dubious information. Many were cloistered in naïve, self-reinforcing, insular echo-chambers. This had consequences for experts. US Naval War College Professor Tom Nichols had been warning of a "death of expertise" since 2014:

I wonder if we are witnessing the "death of expertise": a Google-fueled, Wikipedia-based, blog-sodden collapse of any division between students and teachers, knowers and wonderers, or even between those of any achievement in an area and those with none at all. ... What I fear has died is any acknowledgement of expertise as anything that should alter our thoughts or change the way we live. A fair number of Americans now seem to reject the notion that one person is more likely to be right about something, due to education, experience, or other attributes of achievement, than any other. ... We now live in a world where the perverse effect of the death of expertise is that, without real experts, everyone is an expert on everything. ... There are no longer any gatekeepers: the journals and op-ed pages that were once strictly edited have been drowned under the weight of self-publishable blogs (like, say, this one).[53]

University experts faced their own crises across the West, too. Top scientists worried they "wouldn't be productive enough for today's academic system." Physicist Peter Higgs wondered whether "work like Higgs boson identification" would even be "achievable now as academics are expected to keep churning out papers."[54] Computer scientist Cyril Labbé derided the accelerating publish-or-perish "spamming war" at the heart of science.[55] Benjamin Ginsburg's 2011 *The Fall of the Faculty* showed how non-academic administrators increasingly seized control from scholars with PhDs. Many colleagues were fed up with university research audits.

They were tired of how scholarly assessment metrics valued the quantity of their publications over the quality of their findings. Journalist Richard Harris's book *Rigor Mortis* showed how an overemphasis on productivity at the expense of "getting the right answers" not only encouraged shoddy biomedical research; it also discouraged scientists from investing time in replicating or verifying published studies' results.[56] Anthropologist David Graeber announced the "death of the university" as a "guild of self-organized scholars." Graeber pointed to academia's paperwork-riddled bureaucracy and the subordination of scholars to corporatized university "managerial feudalism."[57] Enrollments dropped and tuition costs rose. Some asked whether we are seeing "peak higher education."[58] Amid all of this, many PhDs in academia found themselves underemployed. They joined the armies of adjunct teaching staff upon which universities increasingly rely. These adjuncts often have small part-time salaries, little job security, and no employment benefits.

The 2016 US election's protest politics channeled raw anger, disillusionments, and disenchantments of sorts that could never be domesticated by the scrupulousness of the Finnish engineer. Donald Trump's charismatic authority contrasted sharply with the measured, calm, detached temperament of Finland's trusted STUK nuclear regulatory experts. Rapid publish-or-perish academic research tempos blazed past the plodding pace of Posiva's plans to deliver an updated version of the Safety Case every ten or fifteen years until the decommissioning of the repository around 2120. America's deflated faith in experts' capacities to anticipate even quite short-term futures contrasted with the Promethean ambitions of Safety Case experts' multimillennial projections. It made them seem anachronistic or far too trusting. As Paypal founder Peter Thiel said in 2014, America was growing "very hostile to big ideas," as projects like "the Apollo program are quite unthinkable today." I began to appreciate how Safety Case deep time reckoning was about putting the fearful politics of denial aside and dreaming big. It was about imagining better tomorrows, embarking upon ambitious technological projects, and embracing open futures. This spirit of inquiry, I realized, can offer vital lessons as some asked, "Has America lost its ability to dream big?"[59]

So, ultimately, this book's reckonings have the most to offer readers in places where the deflation of expertise has become the most engrained.

In the United States, for example, the Trump administration recently advanced a proposal to reform its National Environmental Policy Act regulations to read: "Effects should not be considered significant if they are remote in time, geographically remote, or the product of a lengthy causal chain."[60] In contexts like these, Finland's Safety Case experts' efforts to discern deep time seem truly alien. To fully grasp these contrasts, though, we must first learn more about twenty-first-century Finnish society. We must explore which aspects of Finnish culture helped lay the groundwork for Safety Case experts to make bold journeys across time without collapsing into self-doubt or caving to onslaughts of public criticism. Posiva's projects to reckon deep time were, after all, enabled by a Nordic country known for its trust in governance by networks of highly trained experts. Many (but certainly not all) Finns placed faith in the authority of expertise, education, and technology. As the Anthropocene and the deflation of expertise simmer, this openness to expertise-driven progress can be held up as a counterpoint to today's globalizing ecological and intellectual crises. It can help us imagine our worlds differently.

However, we should not idealize Finland. Not all Finns embraced expertise. Among those who did, the embrace had both promises and perils. Many of the Finns I met were delightfully well-read on matters ranging from urban planning to industrial design, aerospace technology to climate change. Many spoke three or more languages. Yet more than a few disproportionately admired engineers and natural scientists far more than, say, literature scholars or philosophy professors. Many Finns, perhaps rightly, felt that their respected domestic scientists, ministry technocrats, and regulatory experts had earned the citizens' trust. Yet some tended toward blind deference to authorities and uncritical faith in scientific realism. Not all aspects of Finnish expertise culture can, or should, be adopted elsewhere. Trusting in technocrats can be very dangerous in places prone to corruption, dishonest officials, or incompetent scientists. In Finland's finer moments, though, much of its population tended toward a grounded and pragmatic respect for rigorous expert inquiry. Studying this anthropologically can help us think beyond today's intellectual and ecological crises. In this spirit, let's explore how Finnish expertise culture often diverged from—but, unfortunately, sometimes succumbed to—the deflation of expertise.

TRUSTING TECHNOCRATS

Posiva's long-sighted Safety Case projects did not happen in isolation. Many Finns from many walks of life told me of their country's acceptance of big, centralized, hierarchical institutions. These included nuclear power plants, public transport systems, government ministries, and Finland's welfare state. Some pointed to polls casting Finland as unique in its high levels of trust in its domestic civil servants, experts, police, educators, pilots, engineers, and welfare programs.[61] Others explained how Finland's lack of historical experience with technological disasters like Chernobyl, Bhopal, or the BP Oil Spill had garnered Finnish industry and regulatory institutions heightened public trust. Some pointed to Finland's relatively few political corruption scandals. Still others told me about Erkki Laurila, who worked in Finland's Linkomies Committee from 1958 to 1963. This committee foresaw the Finnish state supporting science and technology in service of the county's national development. Laurila called this outlook "Ilmarinen's Finland," a reference to the smith in Finland's national epic *Kalevala*. Ilmarinen made a magical artifact called Sampo. According to folklore tales, Sampo, a product of technological ingenuity, brought riches and good fortune to its holder. Adeptness in technology, as Laurila saw it, could deliver Finland the same.

Finnish social scientists have described how "Finns value things such as Enlightenment, state, bureaucracy and technology" and "in general count on expertise, technology and authorities."[62] They have explained how a "belief in enlightenment thinking" created conditions in which "science and education" become core values that "characterize and construct national identity." These values have stoked "almost a mania for new technology."[63] They have noted Finland's "trust in technology and its ability to produce welfare," the "strong position of administrative bodies," and the "involvement of academic intellectuals in the creation of the national ideology."[64] They have studied Finland's icon of the "virtually infallible" engineer.[65] International observers have noted Finland's "high level of trust among citizens, a cooperative attitude in society and a sense of solidarity."[66] Oxford's Reuters Institute's 2016 Digital News Report showed that Finns' overall trust in their media outlets was the highest out of twenty-six countries. Sixty-five percent reported trusting domestic news.[67] Of European countries, Finns have been placed sixth for trusting in Parliament, fifth for trusting politicians, and fourth for trusting political parties.[68] This trust has been cast as the key to Finland's world renowned public education system[69] and to how social clout is doled out in Finland.[70] It has been cast as the key to Finns' high levels of participation in voluntary organizations like sports, charity, and cultural associations.[71]

Many Finns I met, expert and lay alike, enjoyed self-characterizing Finnish culture in conversation with me, the anthropologist from afar. Some enjoyed self-stereotyping Finnish ways of life. Many reflected on Finns' strong faith in fact-driven reason, state authorities, and one another. Some said they were more likely to trust career civil servants, researchers, and engineers than, say, certain sophists in Parliament or the attention-seeking "political broilers" heading Finland's ministries. Others lauded the civility of Finnish debate culture's *asiat riitelevät, eivät ihmiset* ("the issues fight, not the people") ethic. Some associated Finnish trustworthiness with speaking in a direct, concise, neutral, no-nonsense way. Others associated it with being punctually on time for meetings and always keeping one's word. Some linked Finns' penchants for *hiljaisuus* ("quietude" or "silence") or being *omissa oloissaan* ("to oneself") to the high value they place on giving each other space for inward contemplation.[72] Some noted Finns' voracious reading habits and aversions to small talk or *tyhjän puhuminen* ("empty talk"). One Estonian told me how he saw Finns as always wanting to be the "perfect student" by doing everything properly, by the book, and with precision. As a Helsinki-based Russian expat recalled a drunk Finn once telling him: the "most important thing about living here is *luottamusperiaate* [the principle of trust]. ... In Finland we trust each other. We trust officials, police, and people."[73]

Finnish trust, cooperativeness, meticulousness, national solidarity, and orderliness have often received credit for Finland's relative openness to nuclear initiatives. In 2011, the Finland Promotion Board website of Finland's Ministry for Foreign Affairs posted an article about a poll conducted shortly after Japan's Fukushima nuclear disaster. It described how, even though public support had slipped, 85 percent of respondents still considered Finland's nuclear power plants "absolutely safe." After asking locals why, the author explained that "the word 'trust' is still heard loud and clear here: trust in the technological capabilities of the nuclear engineers, in the terrain, in the transparency, openness and honesty of the operators and construction companies, as well as of the regulators."[74] Finnish trust was a key image in Finland's national brand. It also appeared in coverage of the Olkiluoto repository. As the BBC reported in 2010, "in Eurajoki in Finland, where the local council decided seven years ago that it would like to see the waste from the country's nuclear reactors buried in its backyard, the T-word is everywhere, nestling alongside its spiritual siblings openness, honesty and transparency."[75] This recalled a 2006 BBC article written by the same journalist after he met with higher-ups from Posiva:

"It boils down basically to trust," comments [Posiva manager] Timo Äikäs. "When you make a decision concerning this kind of thing, which takes us to 2100 when the final sealing takes place, there will always be uncertainty. So you have to have trust." But Timo Äikäs believes his system and his team deserve the trust they have found in Eurajoki, and that Onkalo will prove as safe a resting place for highly active radionuclides as can be found, barring any surprises with the local geology.[76]

But not all were impressed. A frustrated Greek citizen living in Helsinki complained that Finns lack a "poetic predisposition." He criticized their embrace of sober pragmatism and their trust in technical competence, bureaucrats, and rule-governed formality. For him, Finns were ordered, rational, disciplined Apollonians, not spirited, sensual, emotional Dionysians. An expat from Iran complained that Finns' lack of passion left him "tranquilized into a Nordic slumber." A Finnish colleague in the humanities lamented that Finland's consensus-valorizing debate culture made for boring nonadversarial talk radio and failed to foster healthy criticism of society's norms. One social critic derided Finland's outdated "faith in the rational mind of an engineer."[77] Others rolled their eyes at how Finland's high levels of public trust fostered public complacency and blind loyalty. A wisecracking Finnish academic once told me that "in Finland, you get to observe a near total societal embrace of Foucauldian discipline and governmentality." A critically minded architect I met discussed 1980s sociological analyses of Finnish cultural codes, social class, and hierarchies that depicted Finland as the "anti-France." Many Finns, he suggested, embrace naïve realist depictions of life, art, and media without acknowledging cultural subtexts or contexts.

Some Finns criticized their compatriots' trust in nuclear technology. Critical scholars explained how, in the first decade of the 2000s, Finland's Ministry of Economic Affairs and Employment worked in cahoots with powerful people in industry, the media, and Finland's parliament. They built a "pro-nuclear network" and a "hegemonic political discourse" such that "opposition to nuclear power had become stigmatized."[78] Meanwhile, some participated in antinuclear demonstrations like the Olkiluoto Blockades, in which activists blocked roads surrounding the Olkiluoto nuclear power plant. Others developed activist projects like Case Pyhä-joki, a "transdisciplinary artistic expedition" to "explore artistic perspec-tives on the vast changes planned in Pyhäjoki, through the planning of a nuclear power plant at the site."[79] In 2010, Greenpeace placed a "monu-ment of selfishness" in a Helsinki park memorializing names of Parlia-ment members who recently voted for expanding the country's nuclear capacity. Finnish rock band Eppu Normaali's ironic 1980 song "Suomi-ilmiö" ("Finland Syndrome") criticized Finns' culture of nuclear trust. It was written soon after the United States' 1979 Three Mile Island nuclear accident near Harrisburg, Pennsylvania:

Although in Harrisburg one needs to lock one's windows / Finland is always safe / Harrisburg is somewhere on another planet / it could never happen here / beneath the birch and the star / Can perfection exist in any form? / Yeah of course, among other things, at Olkiluoto / None are as smart as an engineer / its perfect / the button and pipe / Uranium splits / and produces the lamp's light / but no other countries other than Finland are free from risk / We have quite a selection of infallibility / [Finland's former President] Kekkonen, a Finnish-Soviet Treaty of Friendship, and [Finnish power company] Imatran Voima / no danger fits in our routines / unless in China the Finland syndrome would occur / Uranium splits / and produces the lamp's light / but no other countries other than Finland are free from risk.

During my fieldwork, Finland's advocates and critics of nuclear energy seemed to agree on something: the Olkiluoto repository's national accep-tance hinged on a widespread trust in expertise, technology, education, and state authorities. The divisive question was whether this trust was enlightened and deserved, or naïve and unwarranted. These conditions helped the Safety Case experts' work remain largely nonpoliticized. They enabled Finland to hold closely to the repository construction schedule it set in the early 1980s. The Not-in-My-Backyard politicking that helped stop the US Yucca Mountain repository in its tracks in 2009 was not strongly present in Finland. Nor were the project stagnations seen over the years in France, the UK, Canada, Germany, Switzerland, or elsewhere.

Most Finns I met saw the Olkiluoto repository as a pragmatic solution to an unfortunate problem. Many saw it as an object of curiosity worth reading up on. A few saw it as a cause of outrage, grievance, or fear. Many had a noticeable aversion to speaking outside the realm of their own competence, declining to opine on nuclear waste. They instead referred me to technical specialists formally authorized to speak on the subject. This all helped clear out the political, intellectual, and logistical space necessary for Posiva's work to persist without imploding amid dissent. It helped insulate Safety Case experts from a degree of self-doubt, agitation, and self-searching that may have been provoked elsewhere.

But much of this was changing. Many worried as Finland's traditional forestry industries (e.g., lumber, paper manufacturing, pulp processing) restructured, as the global reach of its once-booming IT sector led by Nokia retracted, and as key mining and metallurgical companies saw financial turmoil. Finland's Talvivaara mine's wastewater leaks had spread nickel, cadmium, zinc, uranium, and aluminum into Eastern Finland's ecosystems since 2012. This, and the high-profile criminal case which followed it, caused many to wonder whether Finland's domestic extractive industry really worked in their interests. Finland's far-right, nationalist-populist, Euroskeptical party *Perussuomalaiset* ("True Finns") stoked anti-elite sentiments. Its leader Timo Soini entered the country's coalition government as Minister of Foreign Affairs and Deputy Prime Minister. Finns under the age of thirty-five became more skeptical than their elders were of the news media.[80]

Meanwhile, disconcerting elements of the deflation of expertise were on the rise. A scientist at VTT derided how Finland's science today is often diluted by the politics of research funding allocation, the interests of commercial and investment capital, workplace audits of research "deliverables," and pressures to engage more with the public. VTT is Finland's traditionally state-funded, but increasingly privately funded, applied technological research organization. A scientist at Finland's geological survey GTK lamented the decreased ratios of government funding at his institution, and worried the same was happening at VTT, the

Finnish Meteorological Institute, the Finnish Environment Institute, and elsewhere. He spoke somberly of how VTT recently had to lay off two hundred people. He complained about how much time Finland's experts were wasting doing paperwork, not research. He felt for his growing number of colleagues who faced declining job security, working contract to contract and grant to grant. He was frustrated by how many companies relied on short-term subcontracted expert labor rather than signing on lifers for stable, secure, long-term careers. At the same time, humanities colleagues at Finland's universities worried about downsizing and increasingly precarious hiring prospects. By February 2018, University of Helsinki professors demanded salary increases during their first walk-out strike.

Our planetary crisis could grow even more dire if the deflation of expertise continues to proceed alongside Anthropocene destruction. We need to overcome both if we are to survive. The rise of these troubling trends in Finland warrants new calls to action and fresh perspectives on expertise's strengths and limitations. To this end, I will now clarify this book's broader approach to science, technology, and expert knowledge. Doing so will help us sort through the positives and negatives of placing faith in various forms of evidence-based reasoning. This will be our final step in getting oriented before we embark upon the coming chapters' experiments in deep time learning.

TOWARD A GUARDED ENTHUSIASM FOR EXPERTISE

I emerged from fieldwork unapologetically optimistic that human hearts, minds, and technological ingenuities can adapt to more long-sighted ways of life. I remain optimistic that Safety Case experts' spirit of modesty, hard work, pragmatism, and self-reflection can offer guidance. However, my optimism for technology remains more tempered than that of, say, futurists like Ray Kurzweil. Like nearly all of the Safety Case experts I met, I spend very little time heralding "the Singularity"—an overconfident vision of a future in which humans and machines converge. Kurzweil and his acolytes claim this convergence can help us achieve superhuman intelligence, transcend death by stopping aging processes, and upload our minds to computers.[81] That sort of overzealous faith in progress is part of what got us into this Anthropocene mess in the first place.

Scientists can and do make mistakes. As any student of history will tell you, scientists can be wrong—both ethically and factually. Many technologies have had unfortunate, sometimes unforeseen, downsides: nuclear weapons, coal power plants, plastics, Twitter, pesticides, lead pipes, and asbestos insulations are just a few examples. Some of these downsides have been chilling. Industrial technologies aided the Nazis' mass executions. In the 1940s, the US nuclear weapons complex injected unwitting citizens with radioactive elements—including plutonium, uranium, and polonium—to test their effects on the human body. Eugenics movements referenced shoddy science to justify racial hierarchies and forced sterilization programs.

Nevertheless, many still assume scientists have superpowers they simply do not have. Science and technology can never teach us what to value, how to pursue a good life, how to act morally, or how to navigate delicate situations. Logic and rationality alone can never teach us how to love, dream, aspire, care, or appreciate beauty. Scientific research can never provide us with full certainty, either. As physicist and philosopher Thomas Kuhn famously demonstrated, scientific paradigms change over time.[82] Before Copernicus, for instance, the Earth was seen as the center of the universe; after Copernicus, it was the Sun. As another example, while Newton's theory of gravity reigned before the twentieth century, it was later overtaken by Einstein's general theory of relativity. We do not know what new scientific paradigms will take the helm in the future.

Still, to chart viable routes forward for healing our planet, science's ethos of meticulousness must be given special platform. Analysis vetted by highly trained experts has a value that freewheeling podcast rants, talking head television pundits, and impulsive Twitter posts—vetted only by likes, reshares, retweets, and hit counts—do not. I am not advocating for anyone to blindly defer to scientists, technologists, or technocrats, and I am not suggesting that citizens in the United States, Russia, or China should start trusting their own technocrats the way many Finns do. But I am suggesting that, before we form judgments on matters of planetary survival, we first make a point to learn about what informed specialists have to say. After putting in the hard work of hearing them out, the ball is in everyone else's court again to decide how and whether to respond. We can then begin debating whether they are on the right or wrong side of history, whether their innovations will bring human flourishing or collapse, or whether their narrow specializations have prevented them from seeing their actions' wider implications.

So, I am a techno-optimist—but just barely. This may make for a boring social media tweet, but my techno-optimism is proudly measured and conditional. Some problems require scientific solutions, others do not. Tech fixes improve some situations, while making others worse. It is all a matter of context. It was, for example, reasonable when, throughout the Cold War, many social scientists wrote books and articles undercutting overconfident views of science's universality. This scholarship questioned humanity's capacity for total rationality and critiqued how a secretive military-industrial complex pointed to numerical calculations and technical models to justify its own authority. That was, and still is, a smart response to devastating nuclear weapons being controlled by a small but powerful elite who claim legitimacy by pointing to their own rationality. However, this does not mean we should abandon faith in science and rationality altogether. It would be equally naïve to reject *all* nuclear weapons, waste, and energy experts' techniques, ideas, datasets, and methods as hopeless tools of soulless control. At least some of these tools can be repurposed as useful approaches that societies can adopt and then redeploy to serve more fulfilling ends.

The Safety Case experts' Promethean long-termism can provide remedies for the ailments of the Anthropocene and the deflation of expertise. It merits our attention, even when the experts behind it vehemently disagree with one another. Not long before I began fieldwork, researchers from Sweden's Kungliga Tekniska högskolan (Royal Institute of Technology or "KTH") published experimental findings suggesting that copper corrodes in oxygen-free, pure water. If this is true, SKB's and Posiva's spent copper fuel canisters would corrode at rates far faster than initially calculated. Many of my informants were confident that the experiment's setup was flawed, and many scientists I met after leaving Finland concurred. Some, however, worried that, if the small group of scientists did turn out to be correct, then it could be a "show stopper" for Finland's and Sweden's nuclear waste programs. Both countries' nuclear regulators instructed their respective repository programs to investigate the issue. Yet neither deemed the findings convincing enough to halt licensing procedures altogether. Many breathed a sigh of relief in November 2014, when SKB announced that two different research teams had conducted experiments that failed to corroborate the KTH researchers' findings. However, the copper corrosion question reared its head again in January 2018, when Sweden's Environmental Court ruled that further evidence of the KBS-3 canisters' multimillennial corrosion resistance must be provided before it would approve SKB's proposal.

Debates like these can, at times, be frustrating. Yet they are what make evidence-based inquiry great. Scientific debates grow more, not less, robust when educated publics, social scientists, humanities scholars, interest groups, environmental court judges, and other learned people openly participate in them. Disagreement generates new ideas; narrow-mindedly enclosing ourselves in intellectual echo-chambers does not. Nor does engaging solely with people with whom we agree. Closing our minds by assuming the world is just a postmodern abyss of meaninglessness, chaos, and power-politics can never liberate us from Anthropocene darkness. But adopting a spirit of adventurous learning can be enriching. Fieldwork showed me how a subtle, abiding, guarded optimism about technological advancement, intellectual discourse, and scientific scrutiny could—in one specific place, in one specific time—make deep time reckoning more achievable, or at least more accessible, for those who believed in it.

Stewart Brand once asked: "How do we make long-term thinking automatic and common instead of difficult and rare?"[83] My answer is that we can look to Finland's Safety Case experts' practical, pluralistic, grounded, "applied science" of deep time reckoning for some clues, but never all the answers. While this may interest futurologists in the interdisciplinary field of futures studies, my goal is different. I myself do not put forth utopian or dystopian scenarios of possible, probable, preferable, or unfavorable future worlds. Rather, I offer a repertoire of anthropological tools—"reckonings"—as a how-to guide for engaging with others' future visions. My methods may contrast with those of, say, existential risk scholars who reckon future nuclear wars, asteroid impacts, super-volcanoes, pandemic diseases, climate change, robot uprisings, and other cataclysmic events in more arcane ways. Yet we all share a common question: how can we foster human and ecological flourishing across future millennia, despite the many powerful economic, political, cultural, and intellectual forces working against us?

Walking toward these distant horizons, let's turn our backs to the Anthropocene and start thinking longer-term. Let's resist the deflation of expertise by opening our ears to long-sighted experts. In this spirit of long-termist learning, let's chart out our trek into far future Finlands.

CHAPTER 1: HOW TO RIDE ANALOGIES ACROSS DEEP TIME

Chapter 1 explores the Safety Case experts' powerful use of analogy to reckon the Olkiluoto repository's far future conditions. It shows, for example, how they studied a present-day glacial ice sheet in Greenland as an analogue: a stand-in feature used for the sake of comparison. This ice sheet helped them make extrapolations, inferences, and projections about the far future fate of a Finnish glacier during and after the next ice age. The chapter also tracks how Safety Case experts studied a crater lake that resulted from a meteor crashing into the Earth roughly 73 million years ago. The crater became, for them, an analogue for how Finland's landscape might change over the next several ice ages.

Examining how multimillennial analogies are made by nuclear, climate, and space experts during the Anthropocene takes us on an expert-inspired learning-journey into futures and pasts. This journey brings us to far future Finlands, the Roman Empire, ancient China, a future Earth that looks like Mars, Africa during our planet's earliest history, other planets, regions of South Africa in 2030, West Virginia during World War II, and elsewhere. It concludes with five reckonings, each one brainstorming how integrating farsighted flights of analogy into our day-to-day ponderings can sharpen our long-termist intuitions. This means routinely doing intellectual workouts that compare different objects across time and space. These mental exercises, inspired by the Safety Case analogue studies, challenge us to reflect imaginatively on possible similarities and differences between distant past, far future, and present-day worlds. They encourage us to resist the deflation of expertise by investing personal time in seeking out philosophical, scientific, and ecological knowledge and then, if possible, broadcasting it across society.

CHAPTER 2: HOW FAR FUTURE WORLDS SPROUT
FROM SIMPLE REPEATING PATTERNS

Chapter 2 is about learning to see future ecological and geological systems through the technical, disciplined, data-driven eyes of Safety Case computer modeling experts. These experts used the power of pattern to weave together quantitative models—highly technical computer simulations—of far future Finlands. Specifically, they used simple distinctions between input and output to establish a sense of consistency that helped organize their dense jungles of interconnected reports, models, datasets, and scenarios. Studying this anthropologically revealed how, say, a data output from one model could serve as a data input into another model, which could then produce data outputs that fed into, say, three other models as inputs, which then produced outputs of their own, which were each fed into two other models as inputs, and so on.

Embarking on a learning-journey of following these elaborate chains of input and outputs will help us see how radically complex patterns emerged from much simpler ones to make visions of far futures appear. It will also lead us to consider how Finland's future climates may change, glaciers may form, shorelines many change position, landscapes may evolve, and more. The chapter closes with five reckonings, each exploring such questions as: How can we scour our daily routines and patterns for routes into better organizing our own deep time learning? How can basic patterns, found right in front of us, help us overcome the overwhelming feelings of meaningless, awe, mystery, terror, or anxiety that can well up as we learn how to gaze into the abyss of deep time? How can taking a close, anthropological look at experts' work help us counter the deflation of expertise?

CHAPTER 3: HOW TO ZOOM IN AND OUT ON DEEP
TIME FROM DIFFERENT ANGLES

Chapter 3 explores Safety Case experts' skills at toggling back and forth between visions of human, ecological, and geological histories (near and distant) and human, ecological, and geological futures (near and distant). Performing these intellectual journeys across time by zooming in and out across time spans is a crucial Anthropocene skill. Mimicking these experts' intellectual adventurousness, the chapter begins by zooming out from the Safety Case: it approaches the entire project as only one momentary blip within the deeper human and geological history of Finland. Next, it zooms in on the collaborative forces that held Posiva's elaborate Safety Case projects together across weeks, months, decades, and years. After that, it zooms in even further on how Safety Case experts maintained their motivations to endure their work's intellectually taxing day-to-day demands.

From this emerges a sense of how the Safety Case portfolio became fully knowable only when many different kinds of experts were viewing it from many different timescales, angles, levels, and perspectives at once. The chapter closes with five more reckonings, each brainstorming different strategies for getting the shortsighted organizations of today to embrace more sophisticated *multiscale, multiangle,* or *multiperspective sensibilities*. These sensibilities must be widely cultivated if we are to survive the Anthropocene. The reckonings also ask how today's communities of deep time reckoners can achieve greater solidarity—pursuing a shared mission to recalibrate experts' and citizens' relationships with one another and with the Earth's future ecosystems. This will require a societal embrace of expert inquiry that is directly at odds with the deflation of expertise.

CHAPTER 4: HOW TO FACE DEEP TIME EXPERTISE'S MORTALITY

Chapter 4 examines the Safety Case experts' cautionary tales about the consequences that can follow when a beacon of deep time learning dies. It tracks how experts summoned, conjured, or channeled memories of Seppo: a highly talented deceased colleague whose "specter" was said to still haunt their scientific community. For some, the "afterlives" of Seppo's expertise manifested as gaps in knowledge left behind by an expert often reluctant to document how he made Safety Case models. For others, it manifested as anecdotes about Seppo's stubbornness, irritability, and intellectual intensity, as well as his more jovial demeanor during sauna nights, workplace parties, or trips abroad. Still others caught themselves asking "What Would Seppo Do?" while troubleshooting at work.

This chapter argues that lessons gleaned from Seppo's death can be sources of learning as we work toward building more long-sighted societies. Lessons of a similar kind can be gleaned from other expert communities reliant on slow-to-acquire, sophisticated, scarce expert knowledges. The chapter closes with five reckonings, each offering ideas for surmounting the expert replaceability problems that so often accompany rare but essential deep time reckoning specialists. Today's societies, it concludes, must resist the deflation of expertise by embracing an ethic of *predecessor preservation*. This means carefully absorbing, tending to, and disseminating insights from prolific deep time reckoners so their contributions to long-termism do not die off upon their biological deaths. Key long-term thinkers like Seppo, therefore, need to be given both special treatment and additional responsibilities.

CONCLUSION: ESCAPING SHALLOW TIME DISCIPLINE

This book's accumulation of reckonings reaches its peak in the conclusion, where I pull many of them together and set them into motion. The question is how we, as denizens of the Anthropocene constrained by the deflation of expertise, can learn to escape the forces of *shallow time discipline* I laid out earlier on in this chapter. To envisage a society that takes this seriously, I draw on several of chapters 1 through 4's reckonings to undertake two final, speculative thought experiments. The first asks if what we need is a radically new education program that introduces entire populations, from an early age, to humanity's inventory of long-term thinking tools. The second asks how a hypothetical society, educated into greater time-literacy, could be reorganized to more holistically embrace futures-thinking. By taking these alternative worlds as inspirations, the chapter argues, we can better reflect on how to rescue hopes for human flourishing from the Earth's ecological death spiral.

1

HOW TO RIDE ANALOGIES ACROSS DEEP TIME

12020 CE: A solitary farmer looks out over her pasture, surrounded by a green forest of heath trees. She lives in a sparse land once called Finland, on a fertile island plot once called Olkiluoto. The area is an island no longer. What was once a coastal bay is now dotted with small lakes, peat bogs, and mires with white sphagnum mosses and grassy sedge plants. The Eurajoki and Lapijoki Rivers drain out into the sea. When the farmer goes fishing at the lake nearby, she catches pike. She watches a beaver swim about. Sometimes she feels somber. She recalls the freshwater ringed seals that once shared her country before their extinction. She has no idea that, deep beneath her feet, lies an ancestral deposit of copper, iron, clay, and radioactive debris. This is a highly classified secret—leaked to the public several times over the millennia, but now forgotten. Yet even the government's knowledge of the burial site is poor. Most records were destroyed in a global war in the year 3112. It was then that ancient forecasts of the site, found in the 2012 Safety Case report "Complementary Considerations," were lost to history. But the farmer does know the mythical stories of Lohikäärme: a dangerous, flying, salmon-colored venomous snake that kills anyone who dares dig too close to his underground cave. She and the other farmers in the area grow crops of peas, sugar beet, and wheat. They balk at the superstitious fools who tell them the monster living beneath their feet is real.

CRATERS, CORPSES, MUDROCK, AND NAILS

Seventy-three million or so years ago, a meteorite slammed into what is today called South Ostrobothnia, Finland. The serene Lake Lappajärvi now rests in the twenty-three-kilometer-wide crater that was made in the blast's wake. Today, locals enjoy trips to Lappajärvi's Kärnänsaari: an island of rock once melted by a collision from the Cretaceous Period. Canoeing there allows you to brush up against Finland's landscape's deep history. That is why the crater lake caught the attention of Safety Case *natural analogue* researchers. These experts used the power of analogy to make predictions about the Olkiluoto repository's far future fate. They studied prehistoric features in places like Lappajärvi as stand-ins—practical tools for making long-term projections about repository parts, geological features, and environmental conditions. They selected these sites as analogues because they are thought to harbor features similar to those anticipated for Olkiluoto.

Safety Case experts explained to me how Lake Lappajärvi kept much of its distinct shape despite the erosion caused by the advances and retreats of past ice ages' glacial ice sheets. Posiva's reports told of "fairly stable conditions and slow surface processes" at Lappajärvi over millions of years. Taking the lake as an analogue, they argued that Olkiluoto could expect to see only limited movement and erosion of its land across multimillion-year futures. They reasoned that Posiva's repository could hold up reasonably well in similar distant future conditions. Lappajärvi's deep histories were, for Safety Case experts, windows onto the waxing and waning of far future ice ages. For me as an anthropologist, they were tools that could be repurposed to help us learn to refine our own deep time reckoning skills during the Anthropocene.

Safety Case experts also studied natural areas outside Finland. For example, they turned the prehistoric Littleham mudstone in Devon, England, into a place for grappling with deep time. There, copper, encased in sedimentary rock, was preserved for 170 million years without succumbing to major corrosion. Safety Case experts drew analogies between this copper and Posiva's copper nuclear waste canisters. They noted how the Littleham mudstone is even more abrasive to copper than the bentonite clay, to surround Posiva's canisters, would be. The latter, they reasoned, may see even rosier futures. Other Safety Case experts made trips to a large ice sheet near Kangerlussuaq, Greenland, where they conducted analogue research on the permafrost, ice, and groundwater. The idea was that scientific scrutiny of the ice sheet could help nuclear waste repository projects in Finland, Sweden, and Canada more accurately project the characteristics of distant future ice sheets. This, they believed, could provide insights about repository safety.

Other Safety Case experts stayed closer to home, doing Safety Case fieldwork at Southern Finland's own Palmottu uranium deposit. At Palmottu, they studied the underground pathways that groundwater flowed through, assessing how nearby radionuclides had traveled over the years. They searched for evidence of past chemical reactions. Their goal was to figure out how Posiva's nuclear waste's radionuclides may or may not move around in Olkiluoto's underground in futures near and far. The idea was that studying present-day evidence of the past was the key to unlocking insights about far futures.

In every natural analogue study, future conditions are reckoned by drawing comparisons between physical formations across time (drawing inferences about long-term futures with reference to evidence from long-term pasts) and space (making comparisons between regions sometimes thousands of miles apart). Posiva's *archaeological analogue studies* similarly stretched the intellect. Safety Case experts looked to a 2,100-year-old human cadaver in China, surrounded by clay and discovered alongside wood, vegetables, silk, and meat. The experts were interested in the clay's capacities to preserve the body and the artifacts. Posiva reported that Xin Zhui's cadaver was well preserved, with its abdominal organs intact and its skin completely there; some of its joints were still moveable, without it ever being mummified. The vegetables and meat were only partially decomposed because the clay had made an air-tight seal around them. In life, Xin Zhui never knew that he would, in death, become a means for reckoning far futures. Yet, for my informants, he helped demonstrate the power of the repository's bentonite clay buffers to contain radioactive waste canisters for millennia.

Safety Case experts also cited Switzerland's repository program's analogue research on 2,000-year-old ancient Roman iron nails, dug up in present-day Scotland. The Swiss study pointed to how the nails saw limited corrosion across the millennia, despite facing conditions said to be more abrasive than those beneath Olkiluoto. From there, Safety Case experts made extrapolations about the endurance of the cast iron inserts that Posiva planned to use to hold nuclear waste in place inside their disposal canisters.

These Safety Case analogue studies can take us on learning-journeys across time. By taking cues from analogue researchers, we can stretch our minds toward the historical timescales of Scotland's Roman nails, and toward the deep timescales of Lake Lappajärvi. This openness to learning from highly trained specialists is crucial in our struggle against the deflation of expertise. It can help us sync our everyday thinking patterns with the Anthropocene's vast temporal breadth. To this end, this chapter follows a few experts' analogue examples across time. It examines how analogue studies are rife with uncertainties, how their critics try to cut the analogical ties that sustain them, how these studies have been made among global networks of experts, and how they fit into the Safety Case's broader arguments. The chapter then discusses nuclear waste experts' analogue studies in relation to climate experts' and space experts' analogue studies. It closes with five reckonings, derived from the Safety Case experts' work. Each reckoning carves out fresh ways of thinking, judging, and engaging with far futures. To better orient us in embarking on deep time reckoning efforts of our own, the reckonings suggest thought experiments that we can incorporate into our day-to-day lives. But, before we get ahead of ourselves, let's start by following an old cannon into pasts and futures.

CANNONS AND ICE

June 1, 1676: the Battle of Öland was raging. The Swedish Navy was grappling with a Danish-Dutch fleet for control of the southern rim of the Baltic Sea. Amid this bad weather, *Kronan*, Sweden's naval flagship in the region, made a sudden left turn. It was one of the largest warships of its kind. The ship's sails began to take on too much wind. *Kronan* tipped over as water gushed into its gun ports, and it soon lay horizontal on the water. Then an explosion rang out, tearing off a large chunk of the vessel's front side. *Kronan*'s gunpowder storage room was ablaze. The ship, along with around eight hundred men, loads of military equipment, and piles of valuable coins, then sank to the bottom of the sea—eighty-five feet down. Sweden lost the battle. From 1679 to 1686, Swedish divers used diving bells to recover over sixty cannons from the wreck. After that, *Kronan*'s precise location was forgotten. The ship was left in its watery resting place for almost three centuries.

In 1980, a team located the old warship once again. Since then, over 30,000 artifacts have been retrieved from it, in what has been one of the most elaborate archaeological projects in Sweden's history. Safety Case experts now see a bronze cannon from the *Kronan* as an analogue that can help them assess how and whether Posiva's copper nuclear waste canisters will corrode over the long term. My informants explained that the artifact offers insights into (a) how the Baltic's abrasive seawater affected the 96.3 percent copper that made up the cannon's bronze and (b) whether areas of the cannon that had been encased in seafloor clay for centuries were shielded from corrosion. The latter inquiry, they hoped, would shed light on whether Finland's and Sweden's repositories' bentonite buffers would effectively insulate the waste canisters from the underground environment.

The *Kronan* cannon can inspire long-term thinking, and help us to stretch our intellects and imaginations forward and backward across time. But we must always remember to never take analogue studies *too* seriously. They have many limitations. Anders, a Swedish expert with a PhD in science policy and a master's degree in engineering and physics, repeatedly cautioned me about this.

Anders was openly critical of Sweden's and Finland's nuclear waste disposal efforts; that was, after all, his job. He worked for an NGO with a mission to provide Sweden's repository licensing process with independent expert criticism. Anders's NGO was, in fact, funded by the same industry-financed and state-run Nuclear Waste Fund that paid the salary of the Swedish nuclear waste repository experts he set out to criticize. Anders was talkative, opinionated, and articulate. He lauded Sweden's nuclear regulator SSM's commitment to Sweden's constitution's *offentlighetsprincipen*. This "Principle of Access to Official Information" fostered transparency in Swedish government organizations. But Anders critiqued how these rules did not apply to private companies. He was unsatisfied with the amount of documentation that Sweden's nuclear waste management company SKB was making available to him. Anders worried that, when it came to nuclear waste decisions, many Finns were too quick to defer to paternalistic authorities and "men of action" without fully digesting the facts. He worried that they took a "Wild West" approach to nuclear waste decisions—acting first and asking questions later. Anders vented his frustration that elder nuclear industry insiders have been propagating "myths" to overzealous, impressionable, young industry recruits. These myths pertained to nuclear energy's costs, new Generation IV reactor designs, and how nuclear plants fit into sustainable energy systems.

When I brought up analogue studies, Anders emphasized that Finland's Posiva and Sweden's SKB selectively cherry-picked natural and archaeological analogue examples to support their own agendas; they spent little to no time, he said, scouring the globe for counterexamples that could call the KBS-3 repository design into question. Anders's aim was to reveal a devil in analogue research's details. When encountering one, his tendency was to highlight gaps in similarity between the two objects being analogized. He aimed to sever connections between them by pointing to differences between them. For example, Anders saw real limits to what a *bronze* cannon submerged in the *sea* for *centuries* can really tell us about a *copper* nuclear waste canister buried in *granite bedrock* for *millennia*. As another example, he questioned whether a present-day ice sheet in Greenland could really be accepted as a valid analogue for an ice sheet during some future ice age in Finland. The latter, after all, will see eventually far colder temperatures than the former does today. A few months later, I broached Anders's critiques with a Greenland Analogue Project expert from Finland named Aapo. He replied:

It is always this way with geology: you can look at two features just 100m apart from one another and see them as totally different areas geologically. It is a scale issue ... you can scale up bigger and bigger—but of course you just lose and lose detail when you go to a bigger scale. ... But look: you can actually get some real data from ice sheets in Greenland and that is the issue we have to look at. You *need* to go somewhere else with an ice sheet to really do this research and finding an analogue is the best available way of doing it because there are no glaciers in Olkiluoto. Of course, you have certain features in Greenland that you cannot match with the Swedish and Finnish site: for example, the rocks are not exactly the same, but they are very similar. They are in the same age range and they are crystalline granite. But any project in geology involves some measure of extrapolation or of application of your own preexisting expertise and knowledge about the basics of geology.

Aapo's defense of the Safety Case experts' analogies was subtle. First, he affirmed the accuracy of Anders's epistemological critique—his criticism of what nuclear waste analogue researchers were accepting as knowledge. Yet he downplayed the importance of Anders's conclusion. He did this by portraying Anders's emphasis on the differences between far future Finland's ice sheet and present-day Greenland's ice sheet as taking advantage of an ordinary point of dispute already well known among geologists. Specifically, Aapo saw Anders as hyping up common but unremarkable gray areas about how to "scale" one's interpretations of geological features. After that, Aapo simply brushed Anders's critique aside by calling attention to the practical necessity of collecting hard data. He saw analogue experts' efforts as crucial even if achieving perfect knowledge about Olkiluoto's far future is ultimately impossible. Reversing Anders's disanalogizing approach, Aapo then downplayed key differences between Greenland today and Finland tomorrow. He emphasized key similarities between them—namely, those concerning rock age and rock type.

So, Anders tried to unwind an analogical tie by offering an epistemological critique of the Safety Case experts' knowledge. Aapo responded with trying to retighten the analogical tie by appealing to the pragmatic spirit of applied research. Toggling back and forth between seemingly incompatible deep time reckoning viewpoints like these can enrich how we approach long-termist knowledge ourselves.

Aapo's and Anders's disagreements can, for instance, teach us about intellectual optimism versus intellectual pessimism. According to philosopher of science Adrian Currie, this is largely a matter of personal attitude.[1] Optimists, seeing the glass as half-full, tend to highlight the positive aspects of a situation; pessimists, seeing the glass as half-empty, tend to highlight the negative. Both, however, agree on the details: they agree that the glass contains 50 percent liquid. In other words, they do not have an *empirical* dispute. Anders may have (pessimistically) tried to shoot down Aapo's (naïve) optimism by cutting his analogical ties. Aapo may have (optimistically) tried to shoot down Anders's (paralyzing) pessimism by tying the analogies together again. Yet the two did not disagree on basic details about how glaciers work, the *Kronan* cannon's composition, or how Posiva's canisters would be manufactured. Rather, their disagreements lay in how comfortable they were making analogies between different objects. They disagreed about what conclusions they could draw from the analogies and how much weight they ought to be given. To use Currie's terminology, Anders and Aapo disagreed about an epistemic situation: the challenges scientists face when forging knowledge in a given context. They also disagreed about analogues' roles as "epistemic resources": whether the analogies qualified as valid "knowledge, capacities, sources of evidence, and techniques" to solve scientific problems.[2]

That said, Aapo and Anders both underscored how trying to be honest with oneself and others about one's limitations is the first step in becoming a more nuanced, careful, and modest deep time reckoner. Many Safety Case experts agreed. The biosphere report's "Knowledge Quality Assessment," for example, aimed to increase confidence in Posiva's models by openly admitting to the "large uncertainties in the knowledge base." It sought to detail its "assumptions and uncertainties" in a "systematic and comprehensive manner."[3] Still, Anders's criticism was the one that most strongly called us to ask: at what point do the dissimilarities between the two objects in an analogue study become so pronounced that they weaken the analogy?

There will, of course, always be differences between the objects being compared. The *Kronan* cannon was, after all, only said to be an analogue of the canister, not a perfect copy of it. But when do these dissimilarities become significant enough to render them no longer able to shed meaningful light on one another? What degree of sameness, and what kinds of sameness, must be present to make or break an analogical connection? Even if an analogy is imperfect, can it not still add value if at least *some* useful information can be extracted from it? Chatting with Anders showed me how asking critical questions like these is key to moving forward with deep time learning—so long as we can hold onto the optimistic belief, held deeply by Aapo, that evidence-driven reason can ultimately help refine how we envision far futures.

These analogue studies also required imagination. They required a knack for creatively making connections between far-flung locales and far future Olkiluotos. But this was not the seemingly individualistic imagination of a solitary daydreamer, an artist at an easel painting solo, or a novelist staying up late writing alone. Rather, it arose from interactions between different kinds of experts, landscapes, published findings, future visions, institutions, scientific techniques, corporations, and technologies.

These networks were global in scope. The Natural Analogue Working Group (NAWG), originally founded under the auspices of the European Commission, has, since 1985, held fifteen international workshops exploring how analogue studies can support toxic waste disposal projects. Its members hail from India, Germany, Japan, the UK, the Netherlands, France, the Czech Republic, South Africa, the Slovak Republic, Israel, Poland, Australia, Romania, the United States, South Korea, Sweden, Canada, Croatia, and Taiwan. NAnet, another program also funded by the European Commission, brought experts together from ten repository initiatives to promote analogue studies. In both networks, analogical connections made between distant futures and pasts sprouted from professional connections made between experts across national borders.

As the deflation of expertise closes minds worldwide, this analogue research's spirit of global exchange merits appreciation from publics, politicians, and social and news media. It must be elevated from its current obscurity and popularized; it must be broadcast widely, helping others resist Anthropocene shortsightedness. Analogue research must not be drowned out by today's widespread deflations of optimism, trust, and confidence in international expert collaboration. To more fully embrace its long-termism, we must place a guarded faith in the promises of scientific knowledge interchange within and across national borders. This means celebrating the power of globalism to deliver humanity important analogical tools for reckoning futures both near and deep.

That said, analogical reasoning itself is nothing new. Its centrality to scientific argumentation is hardly unique to nuclear waste experts. Analogy is at the very heart of geology's "uniformitarian" methods of extrapolating about past phenomena by making analogies with phenomena observable today. A geologist, for instance, might look at present-day places like the Bahamas to understand how limestone formed in New York State hundreds of millions of years ago. Climate scientists might look at past climate change processes to predict present and future global warming episodes. Medical researchers often study mice and other evolutionary cousins of *Homo sapiens* as analogical stand-ins for humans in experimental trials.

Then there are more specialized examples of the power of analogy. The Consultative Group for International Agricultural Research (CGIAR)'s Research Program on Climate Change, Agriculture, and Food Security has developed a climate analogues website. It was funded by the European Union, the United States, Thailand, and Switzerland, plus other governments and private donors. The online tool looks at rainfall and climate forecasts for future places across the Earth. Then it makes analogical connections between them and present-day regions already seeing those conditions. CGIAR's idea is that if, for example, the climate of Durban, South Africa, in 2030 will resemble that of northern Argentina today, then the maize farmers in the latter region should be in contact with their counterparts in the former region. They could, perhaps, teach them how to adapt to coming climate changes. As another example, in recent years NAWG has also discussed the use of analogues for carbon sequestration proposals.

But analogue research is not just global but also interplanetary in scope. NASA scientists have studied deserts in Arizona as analogues for evaluating how robotic systems and vehicles will fare during future Mars missions. In 1969, they sent astronauts to Craters of the Moon National Monument in Idaho to work with lava fields and volcanic geology in preparation for the Apollo 14 Moon mission. NASA biologists have studied freshwater microbialites in British Columbia's Pavilion Lake. Microbialites are tall underwater structures formed by the activity of microbes. They have served, for the biologists, as analogues for the earliest known remnants of life on Earth, and for other planets' capacities to support life. Scientists have studied archaebacteria living in Earth's deep-sea hydrothermal vents as analogues for microbes that could, speculatively, be found on Mars.[4] For many experts, analogues are practical tools for teaching the mind how to stretch across time and space. But this does not mean analogues alone are enough to reckon far futures.

MULTIPLE LINES OF REASONING

The most powerful scientific views of the past and future are methodologically omnivorous: they combine many different epistemic resources together into complex models of very local phenomena.[5] These resources can be anything from computer simulations, to artifacts left behind from the past, to analogue arguments. Analogues work best when scientists draw from multiple examples simultaneously. They work best when engaging their object of study at many "different levels of analysis," with each analogue performing "different roles, compensating for each other's failings."[6] It would be an exaggeration to claim that Safety Case experts achieved these ideals, but they did have a sense that analogues should be corroborated and bolstered by other forms of scientific evidence.[7]

Safety Case experts strategically approached distant futures through what they called "multiple lines of reasoning." This meant intentionally having many different teams of experts, each with different disciplinary backgrounds and intellectual tendencies, working in parallel on the same long-sighted challenges. The rationale was simple. A team of, say, metallurgists and engineers may have weaknesses that can be compensated for by, say, a team of geologists and biologists' strengths, and vice versa. Incompatible views were taken to be not only competing with one another, but also complementing one another. Together they painted more vivid portraits of future worlds by viewing them simultaneously from multiple angles. Locating analogue studies within these reasoning lines can help us understand their position in Posiva's wider project.

The Safety Case portfolio's main line of evidence was its engineering descriptions of the repository's "barrier system." This line described how Olkiluoto's granite bedrock and the repository's copper, clay, and iron components would contain Finland's waste given Olkiluoto's underground chemical, hydrological, and geological conditions. It also considered the barrier effects of the tricky-to-dissolve ceramic materials inside the pellets that made up Finland's spent fuel bundles. Then there was the portfolio's "Performance Assessment" section, which included number-crunching models. That was the Safety Case's primary systematic analysis of how the repository's mechanical parts, heat levels, nearby groundwaters, and so on may interact over hundreds of thousands of years. Alongside the Performance Assessment were less-likely scenarios modeling potential future repository problems, breakdowns, and "lines of evolution." There were also hypothetical "release scenarios," in which radionuclides enter ecosystems on or near the Earth's surface. The Safety

Case's "Safety Assessment" section estimated annual radionuclide doses to human populations, as well as those absorbed by regional plants and animals. It assessed a variety of scenarios of varying plausibility. Posiva considered these estimations of radiological consequences in light of nuclear regulator STUK's YVL regulatory rule guides. Posiva's Safety Case report then determined, as nearly all stakeholders expected it would, that its own Olkiluoto repository would pass regulatory muster. It was, however, ultimately up to STUK's technical reviewers to officially accept or dispute the Safety Case's determinations. Then it was up to Finland's Ministry of Economic Affairs and Employment to formally issue a repository construction license.

So, Safety Case experts entertained several different forms of scientific evidence. They did not, however, give them all equal weight. Alongside all these engineering, modeling, and quantitative lines of reasoning was a report called *Complementary Considerations*, a hodgepodge of public relations information and qualitative evidence made to persuade wider audiences of the repository's strengths. This is where many of the Safety Case experts' analogue studies were found. *Complementary Considerations* was presented as filling gaps in knowledge that computer modeling and engineering calculations alone could not fill. Most of my informants saw analogues as playing a "supporting" role in the Safety Case's hierarchy of knowledges. To them, analogues were a less-highly regarded backup line of reasoning, there to step in only when more quantitative lines broke down. As an anthropologist, I often felt too much faith was placed in quantitative models and too little in empirical, qualitative analogue studies. Certain geologists agreed with me. The concreteness of the natural and archaeological analogues was, to them, more persuasive than Posiva's quantitative models' many labyrinthine layers of abstraction (which will be the focus of the next chapter).

Most informants, however, agreed with me on something broader: something like the Safety Case experts' "multiple lines of reasoning" principle can provide a compass for navigating far futures. The Safety Case strove to take higher-resolution pictures of Olkiluoto's far future by offering evidence, arguments, and calculations targeted at various audiences with varying degrees of faith in the various reasoning lines it presented. This multipronged approach can help a deep time reckoner appreciate the power of what one informant called "strategic redundancy." This meant following an "Even if X, then Y" logic. *Even if* one does not have faith in Safety Case modelers' simulations of future Finnish ecosystems and hydrogeology, *then* one can still turn to, say, the engineers' defenses of the repository's barrier system or reports on "mechanical strength" tests done on Posiva's copper canisters. As another example, *even if* one does not trust those engineering reports, *then* one can still turn to the qualitative prose scenarios describing Finland's very far future or to natural or archaeological analogues supporting the KBS-3 design. This reminds us that, when we reckon deep time, analogues must serve as but one method in a more holistic, multifaceted approach. All the better if this approach is methodologically omnivorous. All the better if it combines several different methods into something greater than the sum of their parts.

We have now learned to put analogue studies in their place. We have positioned them within the Safety Case's broader layout of coexisting reasoning lines. Next, we can get more specific: we can learn how to more usefully integrate analogical reasoning into our own long-termist lines of reasoning. This will be useful to us during the Anthropocene and the deflation of expertise: a tumultuous time, when all futurological knowledge, and all lines of reasoning, can potentially have value. Analogues can help us learn to learn more and more about our planet's yesterdays and tomorrows.

TRIPS ACROSS TIME

Historian Gabrielle Hecht has approached nuclear waste natural analogues as what she calls *interscalar vehicles*, objects that can draw scholars and others to "move simultaneously through deep time and human time, through geological space and political space." They can be a "means of connecting stories and scales usually kept apart."[8] This means first learning to follow an analogue across time and place, as I did with the *Kronan* cannon. Hecht's example was the Oklo natural fission reactor in Mounana, Gabon, Africa. Oklo is a uranium deposit that, long ago, had its own self-sustaining nuclear chain reactions for a few hundred thousand years. Today, nuclear waste programs point to the radionuclides that the fossilized reactor left behind as analogues for how far future radionuclides at nuclear waste repositories may or may not disperse. They emphasize that the radionuclides did not disperse far from the Oklo site. Some critics, like Anders, saw the Oklo analogue as dubious.

When Hecht followed the Mounana interscalar vehicle across time, she found herself engaging with the theories of an eccentric scientist named Paul Kuroda on our solar system's origins, with the Oklo deposit's fission reactions' origins 1.7 billion years ago, with France's concerning history of twentieth-century uranium extraction in postcolonial Africa, with France's post–World War II nuclear weapons and energy programs, and with the ways Mounana locals are negatively affected by uranium mine tailings today. Taking this learning-journey across timescales, for Hecht, brought "industrial time into dialogue with deep time, bodily temporality into dialogue with planetary temporality."[9] This, to me, underscores that we do not have to be physical or natural scientists for analogues to nudge us toward reflecting more imaginatively on possible similarities and differences between distant past, far-future, and present-day worlds.

One does not need a PhD—or to be an anthropologist like me, a historian like Hecht, or a nuclear waste scientist like Risto—for analogues to widen the time horizons of one's thought. Journalist Jonathan Jones has, for one, looked at NASA's Curiosity Rover's pictures of barren lifeless Martian landscapes to speculate about what Earth's far future landscapes might look like if environmental degradation destroys life here. To prepare for World War II fighting in Northern Italy, US Army personnel first trained in the crags of West Virginia's Seneca Rocks. They saw the crags as analogues for those found in the Dolomites mountain range overseas. Learning to integrate flights of analogy like these into our day-to-day ponderings can stretch our intellects across time and space too. This is a crucial task during the Anthropocene—a time when the "boundary between the ancient and the contemporary" can appear "mucked up."[10] It means trying to be more like a Finn who reads about Lake Lappajärvi in this book, on the website Atlas Obscura, or on Posiva's online Databank and then decides to go paddle-boating. On the water, he or she starts pondering distant past asteroid impacts, multiple ice ages' erosion processes, and far future nuclear waste repositories.

To resist the deflation of expertise, we can all study up on the publicly available findings of analogue experts who work on projects of nuclear waste disposal, space exploration, climate change, and carbon sequestration. Doing so, one may ultimately become only an amateur analogizer: a dilettante who never quite attains the skill of a highly trained expert who conducts analogue research professionally. That is fine. What matters is that we all work to improve our long-termist skillsets, even in incremental ways, while extending the societal impact of today's most long-sighted experts. Our efforts to acquire scientific knowledge must, however, be supplemented with imagination. When looking at, say, an ancient artifact in a history museum, we could try to start thinking analogically. Why not imagine how analogous appliances found in homes today might be presented in a history museum thousands of years from now? This exercise turns a present-day exhibit into an analogue for a far future one. Or, when watching a documentary on excavations of ancient Mesopotamian settlements, why not try to imagine what present-day Seoul, Buenos Aires, Helsinki, or Mumbai might look like to archaeologists excavating them millennia from now?

Combining expert-driven learning with futurological exercises like these can become a kind of long-termist intellectual calisthenics. A person in Bangladesh, New York, Rio de Janeiro, Osaka, or Shanghai could, for instance, try to imagine their area submerged by, or fighting off, future rises in sea level. When taking a bus on Maryland's Interstate 68, one could look at the shaded rock layers often found where rural highways are carved into hills. One could reflect on how the strata indicate ways that the local landscape has changed over hundreds of millions of years.[11] Then one could ask: What might this area have looked like in each of the past time periods that each rock layer represents? If I were a stratigrapher living in the far future, would I see "technofossils" of twentieth-century human activity—the distinct human-altered strata layers that mark the proposed Anthropocene epoch?[12]

Long-termist expertise can sharpen our abilities to answer these questions. Analogue studies can provide resources for more accurately visualizing the present-day world in radically different states across time and space. Scientific knowledge can add sophistication to these futurological thought experiments, which can be attempted by anyone anywhere. The key is to (a) make a connection between our immediate surroundings and a dramatically long-term future or past and then (b) try to envision it as accurately as possible by drawing, analogically, from information and imageries we already have in our heads of concrete, real-world locales out there today. Doing this can outfit us with an evidence-driven toolkit more tangible than navel-gazing speculation. If entire populations were to commit to developing these skills en masse, a much grander transformation could take place. Contemporary societies could inch closer to liberation from their dangerously shallow time horizons.

As the Anthropocene and the deflation of expertise take hold, learning to better reckon deep time becomes a practical necessity. This is as true for us as it is for analogue studies experts. To this end, I close with five reckonings that walk through how anyone can do analogical thought experiments to cultivate their long-termism. Each has been inspired by the learning-journey that analogue studies just took us on. This analogical trek took pit stops in far future Finlands, the Roman Empire, ancient China, a future Earth that looks like Mars, Africa during our planet's earliest history, other planets, regions of South Africa in 2030, West Virginia during World War II, and elsewhere. But now it is time for reckoning. If you, the reader, are able to extract more reckonings from all of this, please write them down. These insights will serve as groundwork for building our capacities to reckon deep time as this book progresses.

RECKONINGS

REIMAGINING LANDSCAPES

To better integrate deep time learning into our everyday awareness, we can begin doing analogical thought experiments in which we routinely try to reimagine how outdoor landscapes change across time. This could mean making analogies between how the landscapes may have looked in distant pasts, and how they may look in far futures. As an example, I currently live in Arlington, Virginia, right outside Washington, DC. Sometimes I head west to hike in Appalachia. A quick Google search can reveal how, hundreds of millions of years ago, the region was home to much taller mountains. Some say their elevations rivaled those found in today's Rockies, Alps, or even Himalayas. The challenge, for me, becomes how to reimagine my surroundings as I trek through the hills. So, I draw, analogically, on the images I already have in my head of what taller mountain ranges look like today. These analogies help me stretch the momentary "now" of my hike by enchanting it with a much deeper history and future. They introduce a pragmatic approximation of deep time into my awareness of my surroundings.

This exercise can be done in any outdoor place. Each plot of land has its own distinct past that, with a little background research, can often be uncovered and transformed into exercises in deep time contemplation. Occasionally, we may find that this research has already been done for us, like at the Ice Age National Scenic Trail. This hiking route in Wisconsin is over a thousand miles long. It brands itself as harboring the "finest features of the glacial landscape." The drumlins, eskers, kames, erratics, and kettles found there can "expand your knowledge of Earth's icy past."[13] They can also become resources for imagining what the area may have looked like during and after the previous Ice Age.

So, the next time we find ourselves outside, we should pause for a moment and imagine the nearby landscapes as they could have been in a distant past, or could be in a distant future. We don't need to live near pristine forests, idyllic meadows, or sublime mountain vistas to do this. If you have the time, you could enhance your thought experiment's accuracy by doing a few minutes (or hours) of preliminary research before embarking. Why not expand your repertoire of epistemic resources— learning about your area's landscape history, plus other analogue landscapes found elsewhere today? Developing a hunger for learning about long-term landscape evolution can stop the deflation of expertise from sapping our curiosity. This hunger can be satisfied through personal efforts to read up on scientific findings about nearby ecosystems and far-away analogues. Fortunately, thanks to the internet, more information of this kind is publicly available than ever before.

This knowledge can help us *defamiliarize* our perceptions of this moment some call the Anthropocene. It can get us to take a step back from what is normalized, think critically about our surroundings, and use the power of analogy to reenvision them at different moments in Earth's history. Such outside-the-box thinking can be rewarding. Corporate coaches have recommended taking breaks from familiar thinking patterns to experience the world in new ways that foster creativity and overcome mental blocks.[14] Cognitive scientists have shown that inspiration can be sparked by perceiving "something one has not seen before (but that was probably always there)."[15] Could learning to better emulate analogue studies experts' thinking patterns help lay publics navigate today's dual crises of expertise and ecology? Could it help societies cultivate their intellectual nimbleness across time—enhancing public awareness of local landscape change across decades, centuries, and millennia?

REIMAGINING URBAN AREAS

City-dwellers can also train their minds to envision their surroundings in different time periods. I found an example just blocks from my current workplace: George Washington University's Elliott School of International Affairs in Washington, DC's Foggy Bottom neighborhood. A quick peek at the website Atlas Obscura reveals that, in 1922, excavation crews clearing ground for building the nearby Mayflower Hotel found fossilized bald cypress trees twenty feet below the city's surface. These trees, which lived to be 1,700 years old, grew 100,000 or so years ago. Back then, America's capital city was a literal swamp. Today, four bald cypresses, planted in the mid-1800s, grow in Lafayette Square right near the White House. When I walk by them, the living trees become interscalar vehicles and analogical resources for thinking across time. They provide me with tree imageries I can draw upon when reimagining the nation's capital as a prehistoric swamp.

Analogical materials of a similar sort can be found in Chicago's Garfield Conservatory's Fern Room too. The room gives visitors a "glimpse of what Illinois might have looked like millions of years ago." As the conservatory's website explains, it features an indoor lagoon along with plants from species groups from dinosaur times, 300 million years ago. When in these cities, our minds can repurpose the cypress trees and ferns as analogues for inserting deep time learning into our daily routines. Anyone with an internet connection can discover more examples. If this includes you, maybe a quick Google search will reveal that the city you call home once had a dense rainforest ecosystem at some point in its geological history. If so, you can do analogical exercises when walking, driving, or taking a bus around your area. Ask yourself: what analogical resources do

I already have in my mind—imageries of, say, the Amazon—that I can tap into to reenvision my neighborhood as if it were a rainforest?

We can also reimagine local cityscapes by turning human-made objects and local buildings into interscalar vehicles. When walking by a landfill, we can bask in the dreary uneasiness one can feel knowing that the plastic wastes there may not fully disappear for many human generations. When coming across a coal power plant, we can contemplate how its carbon emissions contribute to the enormity of climate change's long-term impact. When driving by a car factory, we can feel taken aback by the production process' intricacies. We can consider the sheer number of historical events, technological breakthroughs, labor exploitations, and economic trends that had to unfold first before the factory could be built. Our goal is to discover interscalar vehicles locally and then imaginatively ride them across time. This means contemplating the complex chains of events that brought them into existence, plus those that will, in the future, be caused by their existence.

As another example, I recently found myself looking at a Japanese white pine bonsai at the US National Arboretum in Washington, DC. The tree had been alive since the seventeenth century—the beginning of Japan's Edo period. This was a relatively stable time of peace, isolationism, and rigid social codes. It preceded Japan's 1868 Meiji Restoration. Several generations of bonsai masters had already cared for and trained the small pine tree. It had even survived the Hiroshima atomic blast. In 1976, its bonsai master Masaru Yamaki gifted it to the United States to honor the country's two hundredth anniversary. Standing there, I wondered what wisdom the tree would share if it could think and talk. I tried to envision, step by step, the many centuries of history that had unfolded, and that would continue to unfold, around this arboreal interscalar vehicle, from its sprouting to its eventual decomposition.

The bonsai pine, the landfill, the ancient ferns, the coal plant, the cypress trees, the car factory, and the online Amazonian forest images each evoke past and future complexities. These complexities can inspire fascination, curiosity, or even awe, which can have powerful effects on the psyche. A Stanford University study has shown how awe can expand a person's sense of time.[16] States of awe can be "mind- and heart-expanding" too.[17] So, when we defamiliarize our surroundings by making long-termist analogies or following objects across time, we not only lengthen our thinking's time horizons during the Anthropocene; we also broaden our intellectual horizons by finding inspiration in analogue experts' research during the deflation of expertise. The question is: How can we commit to taking this outlook of deep time awe with us, wherever our lives may lead? How can we muster the self-discipline to undertake independent long-termist learning—expanding our repertoires of analogues and inter-scalar vehicles?

SELF-CRITIQUE THAT DOES NOT PARALYZE ACTION

When doing long-sighted analogical and interscalar exercises, we must be critically self-aware of our limitations. This holds for scientists conduct-ing analogue research, as well. Resisting the deflation of expertise does not mean blindly deferring to experts, closing our minds to nonexperts, or pretending that experts never make errors. We must admit from the outset, just as the biosphere experts' "Knowledge Quality Assessment" did, that far futures are marked by incredible uncertainties. Attaining total knowledge of them is impossible. We must accept that experts and laypeople alike have little choice but to make pragmatic simplifications of the future. They must reduce and distill uncertain tomorrows into something more workable, lest their legs buckle before deep time's many unknowns.

Looking far futures and pasts in the eye will evoke what philoso-pher Emmanuel Levinas called *infinition*: a complexity "overflowing the thought that thinks it," or the "overflowing of the idea by its idea-tum."[18] Reaching out to grasp deep time's ungraspable complexity will, in anthropologist Marilyn Strathern's terms, leave "remainders" in analy-sis. These remainders can elicit more analysis, which leave more remain-ders, which can elicit even more analysis, *ad nauseam*.[19] In other words, there is always more of deep time to know. It can never be fully lassoed into place. Yet the payoff of chasing its ever-receding horizons, trudging onward like Risto's Sisyphean ants, is real. To get there, we must first believe, as the Safety Case experts did, that embarking on this path will be more enlightening than not embarking. We must trust that seeking out interscalar vehicles and analogues—despite their very real limitations, as highlighted by Anders—can offer at least some viable, evidence-driven paths forward.

When doing analogical thought exercises, we should ask ourselves: How can I sustain the optimism of intellect, the disciplined attention to evidence, and the faith in reason-driven self-improvement I need to face down deep time's complexities? How can I adopt a constructively self-critical approach to reckoning distant futures without succumbing to indecision, information overload, or analysis-paralysis? When I make predictions, can I use Donald Rumsfeld's famous formula to help me gauge my progress? This means asking: How can I account for the known knowns (what I know I know), the known unknowns (what I know I don't know), and the unknown unknowns (what I don't know I don't know) that accompany any forecast? Have I accepted that my predictions may someday be revealed as rife with unknown knowns (what I once thought I knew but, over time, realized I did not ever know)?[20]

SHARPENING OUR TOOLS

We must commit to growing into more well-informed analogizers of our futures. This book offers only a preliminary step. Future learning-journeys are necessary. The more analogical resources we gather through personal online research, the more sophisticated our deep time reckoning skills become. The more we learn about climate, space, and nuclear waste analogues already identified by scientists, the more interscalar vehicles we have at our fingertips for learning to glide across time. The more time we spend on Posiva Oy's website's public Databank information archive, the more lines of long-termist reasoning we discover. We can reinforce these information gathering efforts by reading long-sighted books like Alan Weisman's 2007 *The World without Us* or Andrew Shrylock and Daniel Lord Smail's 2011 *Deep History: The Architecture of Past and Present.*

In this spirit, try logging onto NAWG's homepage (www.natural-analogues.com) and learning from some of the long-termist analogue work databased in its Library archive. A good place to begin is the website's "NA Overviews" section, which contains reviews and summaries of recent years' state of the art analogue research. After getting a lay of the land, check out the more specialized subsections. These explore the futures of clays, glass, glaciers, cements, archaeological artifacts, and more. Once armed with this arsenal of analogues, one could switch intellectual gears by picking up literary scholar David Farrier's 2019 book *Anthropocene Poetics: Deep Time, Sacrifice Zones and Extinction*, or his 2020 book *Footprints: In Search of Future Fossils*. Then one can begin combining all these ideas, asking: how could Farrier's view of deep time help me think more imaginatively about, say, the clay deposit futures NAWG scientists have studied in Cyprus and the Philippines?

When we add a new analogue to our inventory, we not only get better at making good analogies; we also get better at breaking bad ones. From there, we can begin posing questions like: What are the limits to what *native* copper in *mudrock* in *Devon* can really tell us about *manufactured* copper canisters in *clay* in *Olkiluoto*? To what extent can an analogue help us envision what our home regions might look like tens of thousands, hundreds of thousands, or even millions of years from now? If my hometown will look like a desert, to what extent can I, analogically, draw on mental imageries I already have of Arizona's deserts or of Mars to help me reimagine it devoid of life? Could I draw on science fiction books or movies? If my area will look like Ice Age Finland, could I draw on what I now know of Greenland's Kangerlussuaq ice sheet or of Wisconsin's postglacial terrain? To what extent? Could using CGIAR's online climate analogues tool help me compare my home region's near-future climate to other regions' present-day climates?

MULTIPLE LINES OF REASONING

Doing analogical thought experiments can help us distance ourselves from our time-bound worlds, defamiliarize them, and imagine them afresh. They can help us reposition our everyday lives in broader horizons of time, training our intuitions to adopt the long-termism necessary to think more clearly about climate change, biodiversity, and sustainability during the Anthropocene. But we must remember to perform these personal fact-finding quests and intellectual workouts with a self-questioning attitude. That can help us adopt the cautious, self-reflective, modest optimism about expert inquiry that Safety Case experts had as a foundation for their long-sightedness. If the public, media outlets, and political organizations opened their minds to analogue experts' findings, the cynical deflation of expertise could begin to reverse. Yet we must acknowledge that no one analogy, or even the power of analogy itself, could ever be enough. Analogical reasoning must be just one among—to use the Safety Case experts' buzzword—multiple lines of reasoning.

For the Safety Case experts, the most robust forecasting projects were those in which a variety of groups, using a variety of vocabularies and techniques, all engaged with far future phenomena at once. As a philosophy professor might put it, analogues were necessary, but not sufficient, for reckoning deep time. In this spirit, we must see our analogues as corroborating, competing with, and/or complementing other approaches—as they did in Posiva's *Complementary Considerations* report. Analogues must be positioned in a lively ecosystem of coexisting reasoning lines. This means that when dueling viewpoints—like those of Aapo and Anders—appear irreconcilable, we must not respond by dismissing the entire endeavor as bogus. Instead, in a spirit of adventurous learning, we must ask: How can I see this disagreement as a teaching moment? Can the debate it sparks generate fruitful questions about deep time analogies and disanalogies? Can these questions sharpen my own sophistication when I venture to make or break long-sighted analogies myself?

2

HOW FAR FUTURE WORLDS SPROUT FROM SIMPLE REPEATING PATTERNS

5710 CE: A tired man lounges on a sofa. He lives in a small wooden house in a region once called Eurajoki, Finland. He works at a local medical center. Today is his day off. He's had a long day in the forest. He hunted moose and deer and picked lingonberries, mushrooms, and bilberries. He now sips water, drawn from a village well, from a wooden cup. His husband brings him a dinner plate. On it are fried potatoes, cereal, boiled peas, and beef. All the food came from local farms. The cattle were watered at a nearby river. The crops were watered by irrigation channels flowing from three local lakes. He has no idea that, more than 3,700 years ago, Safety Case biosphere modelers used twenty-first-century computer technologies to reckon everyday situations like his. He does not know that they once named the lakes around him—which formed long after their own deaths—"Liiklanjärvi," "Tankarienjärvi," and "Mäntykarinjärvi." He is unaware of Posiva's ancient determination that technological innovation and cultural habits are nearly impossible to predict even decades in advance. He is unaware that Posiva, in response, instructed its modelers to pragmatically assume that Western Finland's populations' lifestyles, demographic patterns, and nutritional needs will not change much over the next 10,000 years. He does not know the Safety Case experts inserted, into their models' own parameters, the assumption that he and his neighbors would eat only local food. Yet the hunter's life is still entangled with the Safety Case experts' work. If they had been successful, then the vegetables, meat, fruit, and water before him should have just a tiny chance of containing only tiny traces of radionuclides from twentieth-century nuclear power plants.

TAMING DEEP TIME

Gazing into distant futures can evoke feelings of awe, terror, or mystery. In the face of deep time's boundless uncertainties and possibilities, any sense of what comes next can collapse. We may feel dizzy, our motivation deflated. Worst-case-scenario rumination and "what if?" thinking can run haywire. We may feel paralyzed, as our familiar orientation to the world breaks down, and the "individual and the individual's system of relations disappears from view."[1] Telling a human story in the evolutionary history of our species, in the cosmological history of the universe, or in the geophysical history of our planet can make our lives seem instantaneous. When contemplating Big History, we can be overcome by a sense of meaningless. Our Anthropocene visions of tomorrow can be filtered by guilt-ridden despair about resource extraction, privilege, nuclear weapons, population growth, capitalism, biodiversity loss, climate change, future asteroid impacts, pandemic illness, and human extinction. Our optimism about technological innovation, moral progress, human virtue, cultural values, personal responsibility, and expert collaboration can seem hopelessly naïve.

Nevertheless, the Safety Case experts' work to develop quantitative models, numerical forecasts, and computer simulations of far future Finlands has persisted with a steadfast stride since the mid-1980s. This chapter explores how they came to feel more at home in long-term timescales by cutting through deep time's complexities and Anthropocene melancholies. It examines how they drew on familiar patterns and organizational structures to help them simulate very alien future worlds. Imposing these familiarities on far future Finlands provided their thinking with a semblance of orderliness, enabling them to fend off analysis-paralysis and establish a firmer sense of purpose when reckoning deep time. This kind of intellectual boldness must be given a platform—showcased for the world to see—to combat the deflation of expertise.

Reckoning deep time was not always easy for the Safety Case experts. At times, the sheer breadth of their growing library of technical reports felt no less overwhelming than deep time. One recalled asking, a few years before we met, "So who exactly is in charge of this all now anyway?" Some of his colleagues scratched their heads. Over the years, the Safety Case's complexity had expanded to a point at which its experts had become so specialized, its technical reports so numerous, and the sheer number of "cooks in the kitchen" so large that no single human brain could grasp it on its own. Some compared the growing collaboration to an ant colony. They found great mystery in how the project had begun to resemble a "living organism" driven by something like a "group mind" or a "collective consciousness." They quipped that the paths that Posiva's many volumes of reports took while growing in size, number, and detail over the years had begun to take on lives of their own.

This is not to say, however, that the Safety Case was incomprehensible or disorganized; quite the opposite. Each of the Safety Case's specialized reports was understood by some expert or computer somewhere. So were the interconnections that linked them together into a whole portfolio. But this, at the end of the day, was always a collective understanding. While no part of the Safety Case was unknown, nobody alone could know the total package all at once. Some had mastered the Safety Case's big-picture organizational structure. Others had mastered details about individual subsections of it. But nobody simultaneously comprehended, in full detail, both the whole portfolio and all the parts that went into it.

Studying how so many moving parts came together to form Posiva's radically long-sighted Safety Case project gave me a window into the commitments that experts make when their technical knowledge is stretched to its endpoints.[2] It taught me how they used the *power of pattern* to stretch their intellects into the unknown, to order deep time's seemingly endless complexity, and to organize their increasingly large and intricate teams of experts. Guided by a patterned sense of orderliness, the Safety Case experts found the intellectual backbone to reckon deep time. They never fell into a postmodern abyss of hopelessness.

With that as our inspiration, we can embark on a learning-journey of our own. We can learn to approach distant future worlds through the detached, disciplined, data-driven eyes of an important type of Safety

Case expert: the quantitative modeler. The modeler's mission was to make computer simulations of Finland's tomorrows. To ensure that the reckonings we glean from studying Posiva's modeling projects are widely applicable, let's learn to understand them in terms of the broadest patterns of reasoning found at the heart of their work. To accomplish this, we must trek into Posiva's dense jungles of highly technical science and engineering reports.

MODELING MILLENNIA

Most informants agreed that, if a handful of longtime project insiders were to sit in a room, they could, together, wrap their heads around almost all of the Safety Case. This group included the small "SafCa Group" leadership team that oversaw the project's technical work. The SafCa Group consisted of under ten members. It was made up of consultants from Saanio & Riekkola Oy's Long-Term Safety department, Sweden's Facilia AB, SAM Switzerland GmbH, plus Posiva experts. The SafCa Group was advised by a steering group. It was overseen by Posiva's management's research director, who reported to the company's president. The Finnish Research Programme on Nuclear Waste Management (KYT) kept tabs on the Safety Case project, too. The Safety Case was also overseen by Posiva's quality coordinator, plus "requirements management" computer systems like VAHA. VAHA's aim was to make sure that the repository's mission, design strategy, and functioning fit with STUK's regulatory rules and Posiva's many stakeholders' expectations. Some Safety Case knowledge was backed up in research database programs such as the "POTTI," which stored datasets on computer drives. Other information was databased in the "Rock Suitability Criteria" program, which stored data about Olkiluoto's underground rocks and the holes in which Posiva would bury the copper canisters containing Finland's spent nuclear fuel bundles. The Safety Case experts who developed Posiva's *Synthesis* and *Models and Data* reports were tasked with understanding how all these individual studies, reports, and models fit together into a broader portfolio of evidence.

It took a village to make visions of far future Finlands appear. This village was highly trained and outfitted with intricate technological systems. Its members communicated with one another through emails, text messages, in-person meetings, conference calls, phone calls, releases of reports, and conversations over lunch and coffee breaks. Sometimes Safety Case experts traveled together to conferences abroad. They often had STEM graduate degrees from Finnish universities like Lappeenranta University of Technology, Aalto University, or Finland's *ammattikorkeak-oulut* ("vocational" polytechnic or applied sciences colleges). New recruits required years of on-the-job training to achieve specialization in Safety Case expertise. Some were mentored by semiretired experts who stayed active in VTT's or Posiva's projects as consultants or corporate advisors. A Swedish expert jokingly called these "elephant graveyard" or "elder shelf" positions, similar to academia's professor emeritus positions. To ease new recruits' acquisition of *hiljainen tieto*—experience-based "quiet" or "tacit" knowledge that cannot be recorded in technical manuals—the KYT program took cues from the Finnish Research Programme on Nuclear Power Plant Safety (SAFIR)'s "Man, Organisation, and Society" social science studies. To facilitate stable succession across generations of nuclear professionals, the Young Generation Working Group of the Suomen Atomi-teknillinen Seura (the Finnish Nuclear Society) coorganized mixer events with their counterparts in its Seniors Working Group. Nuclear waste management personnel attended these gatherings alongside nuclear power plant personnel.

The Safety Case project's lifespan was intended to last about 135 years. Finland's first safety assessment was the 1985 "Safety Analysis of Disposal of Spent Nuclear Fuel: Normal and Disturbed Evolution Scenarios." It was a report to the Nuclear Waste Commission of Finnish Power Companies—a forerunner to Posiva. Next came TVO-92, TILA-96, and TILA-99. These reports were published in 1992, in 1996, and in 1999 respectively. Posiva's Construction License Application Safety Case was due to regulator STUK in late 2012. Between each of these successive iterations of the safety assessment, lessons learned from previous iterations could be handed over to Posiva's engineers to help fine-tune the repository's design and layout.

After submitting the Construction License Application, Posiva restructured its corporate organization. It then turned to the next iteration of the Safety Case: the Operating License Application version. That Safety Case would focus less on the repository's design and more on its day-to-day waste disposal work. During my fieldwork, Posiva's plan was to submit it around 2018. Now, however, it is scheduled for submission in late 2021. After that, the next major Safety Case will be for Posiva's Decommissioning License Application, due around 2120. That version is to focus on shutting the repository down and cleaning up afterward. With each major application's Safety Case, and in the Safety Case updates due every ten or fifteen years until 2120, STUK expects new and improved analysis. To put this timescale into perspective: in 1850, 135 years before 1985, Finland was a Grand Duchy of the Russian Empire. It has since seen its independence and birth as a nation-state, a vicious civil war, world war, Soviet Union offensives seizing territory that Finland claimed as its own, difficult economic recessions, and the rise and fall of the global mobile phone giant Nokia. Like elsewhere in Europe, 135 straight years of peace, prosperity, and stability would be difficult to come by for Finland. Nonetheless, societal infrastructure steady enough to support the repository project for this long was, and still is, necessary.

The version of the Safety Case I encountered during fieldwork was part of Posiva's Construction License Application. It was a huge collection of reports containing datasets, models, scenarios, descriptions, diagrams, charts, forecasts, maps, documented findings, and much more.[3] It was a corpus of "evidence, analyses and arguments that quantify and substantiate the safety, and the level of expert confidence in the safety, of a geological repository."[4] One of its prime goals was to offer a calculation of the consequences of future radiation releases to future human beings. This was a response to legal requirements. These were defined in Finland's Government Decree 736/2008 and Finnish nuclear regulator STUK's YVL rule guides. Seeking to meet these safety requirements, Posiva concluded that "radiation doses can be assessed, assuming human habits, nutritional needs and metabolism remain unchanged, with sufficient reliability over a period of up to 10,000 years." They concluded that the repository's "safety functions" can be "reasonably assessed up to one million years after repository closure."[5] For Safety Case experts, the far future became a zone of intense number crunching, data collection, and computer simulation. It became a space devoid of the "tropes of the aesthetic sublime" common in media and scholarly depictions of deep time.[6] It became more a site of technical troubleshooting, logistical

organizing, and drab reportage than of apocalyptic dread, cosmic loneliness, overwhelming horror, or existential uneasiness. Much like the
analogue studies we explored in chapter 1, the Safety Case's quantitative
models distilled deep time into something that felt more amenable to
bureaucratic, scientific, management, and regulatory control.

The Safety Case modelers used numbers and scientific methods to
tame deep time's power to confuse. Some made computer models that
simulated how radionuclides could move through Western Finland's far
future bedrock's fractures and crevices, plus the water channels found
deep underground there. Data from Posiva's Onkalo underground laboratory got input into the models. This data distinguished between various rock types and various groundwaters' mineral and chemical features.
Other experts made models of the underground "near-field" areas close to
the repository. They explored pessimistic scenarios in which Posiva's copper nuclear waste canisters could break open or the bentonite clay buffers
surrounding the canisters could erode. They studied whether this could
lead to far future radioactive leakages 450 meters beneath the Earth's
surface. Data from engineering studies of the repository parts' physical
strength got input into their models, as did data from laboratory studies,
conducted in Helsinki, of interactions between the clay and local groundwater. Still other experts made models that simulated how radionuclides,
if they were to escape from the repository in worst-case scenarios, might
get released into Western Finland's landscapes. They modeled how certain radionuclides might then travel around in distant future lakes, rivers,
forests, fields, and bogs. These "biosphere" models drew on data from
ecologists' fieldwork studies, which were conducted outdoors among the
region's plants and animals.

The models of what Posiva deemed the most likely sequence of future
events were found in the Safety Case's Performance Assessment section. This was an analysis of how the repository's mechanical parts, heat
level, groundwater, and so on were expected to interact over the coming
hundreds of thousands of years. Alongside the Performance Assessment
were less-likely scenarios modeling how "incidental deviation" events
might or might not threaten the repository. These included rocks frac-
turing apart underground ("shear" events) and a defective copper canister coming apart at its welds. Presenting a series of "base" (most likely),

"variant" (reasonably likely), and "disturbance" (unlikely) scenarios of possible futures, the Safety Case's safety assessment sections showed how radionuclides could be released into the Earth's surface and near-surface. That could happen if they were to escape and travel through groundwater channels. This provided the Safety Case "Biosphere Assessment" modelers with a basis for calculating hypothetical radiation exposures to humans, plants, and animals. Once those calculations were made, Safety Case reports presented a series of projections about future radiation exposure events. These included a "reference case" (the most likely scenario), "sensitivity cases" (including some negative future "variant" events), and "what-if cases" (unlikely, quite unfortunate scenarios).

Safety Case modelers explored, to use their terms, multiple "lines of evolution." This meant developing multiple versions of how the repository could or could not affect far future Finlands. Different forecasts were weighted with different likelihoods of occurring. These many models, scenarios, and datasets got stitched together and described in writing in Posiva's *Synthesis 2012* report, which contained powerful stories of far future Finlands. As an example, here is a Safety Case forecast telling of ecological and geophysical changes to occur over the next ten millennia:

Over the next 10,000 years ... groundwater flow and chemistry will recover from the disturbances caused by the excavation [of the underground repository]. ... At 1000 to 2000 years after present, the shoreline will have retreated far enough that further changes will not affect the flow rates in the repository. ... In the longer term, major climatic changes are expected. ... Effects include permafrost, glaciation and associated sea-level changes. ... The effects of an ice sheet have also been modelled considering an immobile ice sheet over the whole of Olkiluoto Island for 1000 years, and a retreating ice sheet. ... Successive glacial cycles will impose similar loads as considered during the first glacial cycle.[7]

Tales of distant future changes in Earth's crust, temperatures, sea level, shoreline, and ice age glacial sheets of ice were conveyed in lifeless scientific prose and numerical models. Some of the most alarming, thought-provoking, or attention-grabbing aspects of nuclear futures were disguised by the banal language of bureaucracy. "Making nuclear boring" can, after all, make nuclear experts appear more objective, credible, or apolitical.[8] My fieldwork gave me a lengthy tour of the highly technical, paperwork-ridden, sometimes-quite-tedious insider worlds of a nuclear expert organization. However, over time, I became less and less interested in the boringness of bureaucracy (which, believe me, was alive and well in Finland's nuclear companies, research institutes, and government agencies). Rather, my fieldwork drew me to study something a bit broader: the sheer *ordinariness* of some strikingly simple patterns of thinking, relating, and structuring projects that Safety Case experts followed when weaving their complex datasets, models, and scenarios together into depictions of distant future worlds. These familiar patterns helped the Safety Case remain coherent as it expanded in size and detail in the face of enormous uncertainty.

The patterns of interest to me now are those of *input and output.* During fieldwork, I often found myself asking my informants questions like this: Amid such tremendous organizational and scientific complexity, how did the Safety Case's models of far future worlds continue to grow coherently over the years? Did they follow a guiding inner logic? How did the expanding Safety Case get its feel, or aura, of internal order? Some key Safety Case experts explained to me how *input/output patterns* helped Posiva's farsighted models' many diverse parts come together into more ordered wholes. These logical relationships, I came to realize, made up the Safety Case portfolio's most basic DNA. Input/output was a basic code that, as the portfolio grew, organized its many potential complexities into something more systematic. For us, input/output patterns can serve as a useful launch-off platform, or a familiar and accessible starting-point, for refining our own deep time reckoning skills.

THE INS AND OUTS OF MODELING DISTANT FUTURE WORLDS

My Safety Case informants often made distinctions between input and output to explain how different parts of the Safety Case got linked together into a larger network. They would tell me how, say, a data output from one model could serve as a data input to another model, which could then produce data outputs that fed into, say, three other models as inputs, which then produced outputs of their own, which were each fed into two other models as inputs—and so on. To get a clearer sense of what I mean, let's start with a concrete example of how these input/output patterns worked. The Safety Case's "Biosphere Assessment" models can serve as a useful case study.

Gazing ten millennia into Western Finland's future, Biosphere Assessment modelers used simulation and calculation techniques to forecast ecological and geophysical happenings on and near the Earth's surface. Their models responded to questions like: at what pace will Finland's shoreline continue expanding outward into the Baltic Sea? What happens if forest fires, soil erosion, or floods occur? How and where will lakes, rivers, and forests sprout up, shrink, and grow? What role will climate change play in all this? To make models of these future events, the Biosphere Assessment followed patterns. They adhered to a "logic of slots or internal gaps."[9] That is, for the biosphere model to become complete, it first required other models to be input into it—and for data to be slotted into the information gaps in the model.

Safety Case experts input, or slotted in, five lower-level models into the Biosphere Assessment. One input into it was the "Biosphere Description." That input provided knowledge about the Olkiluoto area's present-day ecosystems and how they had evolved over the past decades, centuries, and millennia. A second input was the "Terrain and Ecosystems Development Model," which simulated Western Finland's water and land formations. A third input was the "Landscape Model," which was a model of how radionuclides might move around in the region's landscapes and waterways in future millennia. A fourth input was the "Radionuclide Transport" model. That input simulated the places at which far future radionuclides could, in unlucky conditions, be released to Earth's surface and then disperse (but only if they were to first escape from the repository and travel upward toward Western Finland's surface through groundwater channels). A fifth input was the "Radionuclide Consequences Analysis," which pursued Posiva's bottom-line, ultimate goal. It calculated the repository's "potential radiological consequences to humans and other biota."[10] The Biosphere Assessment also took into account models of climate changes slated to occur thousands of years from now:

In simulations with low, intermediate and high emissions, the climate at Olkiluoto is projected to be 0 to 5 degrees warmer with a 0 to 20% higher precipitation rate than at present during the next 3000 years (until the calendar year of about 5000). Furthermore, in the low and intermediate emission simulations in which the AMOC did not collapse, the climate at Olkiluoto was projected to stay 0 to 2 degrees warmer with a 0 to 10% higher precipitation rate than at present between the years 5000 and 12000. In the high emissions scenario, in which the AMOC collapses, the climate at Olkiluoto might cool to near the present day climate or 0.5 to 1 degree cooler between the years 5000 and 12000.[11]

These many inputs, as parts, were stitched together to comprise the Biosphere Assessment as a whole. What emerged was a collection of models forecasting Western Finland's surface conditions. Input/output patterns guided other Safety Case modelers as well. As a second example, take the Radionuclide Transport model. Remember: the Radionuclide Transport model was but one input into the Biosphere Assessment model that I just described. For it to be complete, it first had to take in other models' outputs as inputs. One input was the Groundwater Flow model, which simulated how water will move underground near Olkiluoto. Another was a model of Posiva's repository's physical layout and human-made parts. The Radionuclide Transport model could only be complete after these and other inputs got input into it. From there, the modelers could begin making forecasts peering hundreds of thousands of years into the future. Here's an example of one such forecast:

Except for initially defective canisters or those breached due to extensive rock block movements, most canisters are expected to last more than one million years. The design criterion for the corrosion-limited lifetime of a canister in the expected repository conditions shall be at least 100000 years. ... After approximately 250000 years, the [radio]activity remaining in the fuel will be similar to that of a large uranium ore body.[12]

My main point is this: the Biosphere Assessment and Radionuclide Transport models had something very basic in common. They were organized, from top to bottom, in part by input/output patterns. In other words, they both became complete only by taking in evidence, data, and models drawn from many other sources as inputs. Only then could they become finished models, producing useful outputs that provide insights about far future Finlands.

This pattern was shared widely across the Safety Case's models. Some models took in data inputs from scientific observations made at Finland's Onkalo underground laboratory. Others took in inputs from existing international scientific publications. Some took in inputs from

Finland's geological field study sites like the Palmottu natural uranium deposit. Others drew them from the outputs of other Safety Case models. Often, they took in inputs from a combination of these sources. Once a given model was complete, its outputs could then become inputs into still other models. Connecting together relationships of inputs and outputs helped Safety Case experts organize how, to use their terms, lower-level "submodels" fed into "models" which then fed into higher-level "metamodels." As some explained it, lower-level "sub-subsystems" fed into "subsystems" which then fed into higher-level "systems." Put more simply, input/output patterns helped the Safety Case experts establish a sense of consistency that spanned their dense jungles of reports. Input/output patterns became something like a skeleton or a connective tissue that held so many models and datasets together across so many different levels and areas of the portfolio.

The products of all these interconnections were "model chains" and "data chains" of input/output patterns that spanned many Safety Case research projects, teams, models, datasets, and reports.[13] These input/output chains helped the Safety Case's many models hang together into something more unified. They are what made them more than just an overwhelming, chaotic, confusing mess of deep time documentation. On top of this, for an anthropologist like me, input/output chains could also serve as useful compasses for conducting fieldwork. They helped me navigate not only relationships between Safety Case reports but also relationships between Safety Case experts.

DEEP TIME IN CHAINS

To the untrained eye of an outsider, each Safety Case model report could seem like a single, standalone document. For Safety Case insiders, though, each report contained models-within-models-within-models that were deeply interconnected. Parts of models were found in other models, which became parts of other models, which became parts of still other models, ad nauseam. To zoom in or out on any section of the Safety Case's jungles of models, then, was to reveal complex tangles of inputs and outputs. Studying this anthropologically showed the Safety Case's chains of models to be organized almost like a *fractal*: a structure or design in which its many individual parts share the same patterns across all sorts of different scales and levels.

The input/output pattern also helped with the organization of the everyday professional workflows between experts. The title "Biosphere Assessment," for example, referred not only to a bunch of technical documents and modeling reports, but also to the group of Safety Case experts who developed it. The same could be said of "Radionuclide Transport," "Groundwater Flow," and other teams. Working relationships between these teams depended on *handoffs* of information, documents, and models between one another. Once a group finished a model, they would hand it off to another team, which would then input it into their own model, which would then be handed off to still others, who would then input it into a broader-level model, and so on, and so forth. Input/output patterns steered these workflows.

To clarify how this worked, we can return to the example of how the Groundwater Flow model was input into the Radionuclide Transport model, which was then input into the Biosphere Assessment model. In this input/output chain, workplace handoffs of data spreadsheets, paper reports, and PDF files tended to flow *from* the Radionuclide Transport team *toward* the Biosphere Assessment team. In other words, handoffs between teams of Safety Case experts followed the routes laid down by the input/output chains that linked together Safety Case models. Both the models and the input/output patterns that organized them were "relationally produced knowledge and knowledge productive of relationality" between teams of people.[14]

Building on this point, one informant told me how plotting out all of the chains of connections between the inputs and outputs of Posiva's models would produce a "map" of the total Safety Case portfolio. On this map, any expert could place a "You Are Here" sign locating his or her team's position within the Safety Case collaboration's wider jungle of relationships. This gave the portfolio an "all laid out" feel. The layout helped STUK's regulatory experts navigate it when they reviewed it. This was because a reviewer could track *from where* the inputs being fed into a model came from and *to where* the model's outputs went afterward. If a STUK reviewer wanted to zoom in on a particular Safety Case detail— wondering about, say, microbes' far future impacts on the repository's clay buffers—he or she could follow the input/output trail to the answer. This ability to "move around" or "travel" within the Safety Case's jungle of reports made it appear more credible or authoritative to outsiders. It offered the comforts of pattern in the face of deep time's often harrowing complexities.

Sometimes, however, frictions between experts' models—and between modeling experts—arose. Models could glitch or not "talk to one another" correctly, slowing down progress. Some informants were concerned that, if shoddy data were to get input into a model, the accuracy of the model's outputs could be distorted: "Garbage in, garbage out." If one model's accuracy were to be off kilter, it could distort the subsequent models into which the flawed model's outputs got input. Hence, the error could propagate in each successive model along the chain. I once, for instance, heard a story about a radionuclide transport modeling expert named Gustav (we will meet him in chapter 4). He was frustrated because he had to redo his radionuclide transport calculations thanks to an input getting updated earlier on in the modeling chain. Yet, as one modeler reassured me, if a weak bit of data were to make it into a model, this would only sometimes be a problem. Some of the inputs were far more "sensitive" than others. The error's impact on the safety assessment as a whole, in other words, could be negligible.

Errors were not the only things that could propagate across the Safety Case model chain. Informants described a series of small delays accumulating along it over the months and years. Reports on the Olkiluoto repository site's physical description, the engineered layout of the repository, and the Radionuclide Transport model had all been submitted late. These delays caused even more delays. Reports at the chain's end became overdue, in part, because they first needed to be fed by these late reports' outputs before they could be finished.[15] Because of this pattern, experts positioned at the end of the input/output chain were susceptible to pre-deadline time crunches. The Biosphere Assessment modelers bore the weight of accumulated delays. While STUK and Posiva did give them deadline extensions, one biosphere modeler told me this:

We've been discussing this a lot with a lot of biosphere assessment experts around the world. We always end up being the ones that people blame, saying we are late. Remember, the people two years ago were over half a year behind their schedule, but that is forgotten. So, this goes for everybody who is working with a biosphere: the last link of the chain.

One informant compared the Biosphere Assessment modelers' critics to sports critics who scapegoat a hockey goalie for allowing the opposing team to score a goal. Usually, in a hockey game, a few other teammates' defenses would first have to fail for the puck to get into a vulnerable position in which the opposing team could score in the first place. So, blaming only the goalie for the other team scoring—or blaming only the Biosphere Assessment modelers for the entire Safety Case community's delays—is unfair. The moral of the story is this: when placing blame, one should not single out only the person at the end of the chain, ignoring others' prior shortcomings.

Taking an even bigger step back, another informant compared the Safety Case's chain of inputs and outputs to a public bus system. If a bus driver were to lag behind schedule for just twenty seconds to chat with someone at each of the twenty-two stops on his or her route, tiny delays would accumulate. This would leave bus patrons at the final bus stop with an eleven-minute wait (more than a small annoyance in Finland's frigid winter). This bus analogy revealed a *blame chain* that was flowing backward along the Safety Case's model chain. For example, if a Posiva manager were to blame a Biosphere Assessment expert for being late, the blamed expert could try to remove the blame by pointing to how his or her models' inputs—from the Radionuclide Transport modelers, for example—had been handed off late. If a Radionuclide Transport expert then felt blamed by the Biosphere Assessment expert, the expert could then point to how his or her model's inputs—the repository layout report, for example—had been handed off late, and so on.

In these ways, input/output chains not only structured how Safety Case experts linked together reports, datasets, and models; they also endowed their division of workplace responsibilities with order. They shaped their interpersonal squabbles, too. This was a testament to the powerful roles that simple, repeating, guiding patterns like inputs and outputs can play in organizing endeavors to reckon deep time. For us, the question is how basic patterns like inputs and outputs can help us move forward in our own long-termist learning endeavors. As my Safety Case informants knew well, a sense of logical structure can be a useful foundation for navigating

Anthropocene futures. Our goal, though, is not to become full-on computer modeling experts ourselves; let's leave that to those who are highly trained in systems analysis and data science. Our mission, rather, is to take a step back and ask: how can Safety Case modelers' use of input/output patterns inspire us to establish a more productive sense of order in our own day-to-day efforts to reckon deep time? How can expert-driven futurological initiatives as bold as Safety Case modeling projects receive more clout across society during the deflation of expertise?

PATTERNS PROFOUNDLY SIMPLE, YET SIMPLY PROFOUND

Input/output patterns are, of course, not unique to Safety Case modeling. They can just as easily be seen among plumbers, computer scientists, basket weavers, electricians, and countless others. My lungs, for one, take in oxygen as an input and then exhale carbon dioxide as an output. A power plant takes in coal as an input and produces electrical energy as an output. I typed this chapter on a computer reliant on input/output, or "I/O," systems such as a mouse and keyboard (which input signals into it) and a monitor and printer (which output information from it). And so on.

Input/output patterns are so familiar to us that they feel—as anthropologist Roy Wagner once said of both God and money—"somehow mysteriously in front of things, too elemental for easy or ordinary comprehension."[16] They are commonplace patterns of relationship that help organize numerous areas of human life without us ever thinking about them. With this in view, my informants' seemingly unique and historically unprecedented deep time modeling projects begin to seem like just another example, among countless others, of people drawing on simple input/output patterns to organize their lives, their thoughts, and their work. As an anthropologist might say, if "we see present-day cultures as the offspring of past ones, we see new combinations forever being put together out of old cultural elements."[17] In this case, the "old cultural elements" were input/output patterns. The "new combinations" were projects to simulate far future Finlands. To appreciate this is to appreciate how even the most highly technical scientific projects can have, as their foundations, simple organizing patterns found in even the most mundane areas of human life.

Yet patterns are not unique to humans. Patterns can be found in micro-organisms, whirlpools, forests, snails, snowflakes, spiderwebs, stalactites, ant colonies, crystals, honeycombs, and countless other worldly beings and things.[18] Anthropologists have shown how "patterns are harnessed, nurtured, and amplified by life."[19] Patterns organize our worlds even when we pay no conscious attention to them. Our minds have a deep seated "empathy" toward them.[20] This all can get us thinking about how even the most technical scientific finding is, in part, always the result of many familiar patterns mixing with other familiar patterns, to produce other patterns, which mix with still other patterns, to produce still other patterns, ad infinitum. From this angle, Safety Case deep time reckoning begins to look like just another version of much more basic pattern-making and pattern-following processes.

The everyday accessibility of these vital patterns is their great power. After all, we needed no prior technical background in hydrology, systems analysis, or geophysics to follow the chains of connection that input/output helped lay down to organize the Safety Case models. We just needed to take it slow, closely following where the patterns took us. Yet patterns are just as alive in nuclear waste experts' projects to model far futures as they are in brushing our teeth. This simple fact has a profound implication. It means that at least some of the tools we need to achieve a more long-termist worldview can be found right underneath our noses. These tools are alive in the most basic patterns that organize our worlds. Their quiet power can inspire personal thought experiments, intellectual exercises, and mental workouts we can undertake to make distant futures appear more thinkable. To chart out more paths for long-termist learning, I close with five reckonings.

RECKONINGS

DEEP TIME HEURISTICS

This chapter showed how input/output patterns can help us strip away complexity and reveal more basic orderings that organize expert visions of the future. Yet input/output patterns are, by no means, unique to expert thinking. They are what some anthropologists call "extensible." They can be replicated and redeployed across many spheres of life, in all sorts of ways.[21] So, try repeating this line of reasoning elsewhere. Try using it to help navigate other visions of Anthropocene tomorrows. This means seeing "input/output" as a *heuristic device*—a concept "enabling a person to discover or learn something for themselves."

Anyone can draw upon input/output when learning about tricky ecological processes related to, say, the future of climate change. This might mean picking up a science book in a local library. If you have internet access, it might mean logging onto Google or some other online search engine. In any case, understanding climatological concepts usually first requires an understanding of complex chains of future events. Used as a heuristic device, the input/output pattern can help us follow these chains. Let's take the "positive climate feedback loop" known as the "ice-albedo effect" as an example. When reading up on this, one might initially feel overwhelmed with scientific jargon. However, when one carefully applies the input/output pattern to help organize one's thoughts, suddenly the concept starts to feel less daunting. This is a result of the input/output pattern's power to break down complexity into more comprehensible pieces. Here's an example of how this futurological exercise might proceed:

So, climate scientists say that vehicles, factories, and power plants *inputting* greenhouse gases into the air can, as an *output*, cause temperatures to rise in Earth's Arctic regions. When these higher temperatures get *input* into these Arctic areas, this can produce the *output* of melting the sea ice there. When this melting is *input* into the sea ice, this can, as an *output*, shrink the size of that ice's surface. The *output* of the ice's shrinkage is that fewer of the Sun's daily inputs of heat and light into the Earth will get reflected back, or *output*, into space by the ice's reflective white surfaces. This brings us back to the very start of this input/output cycle: even more heat will, as a result, now be *input* into the Arctic sea ice, which has the *output* of starting this input/output chain all over again.

WEAVING TWO-PART PATTERNS TOGETHER

Input/output patterns can help us weave threads of thought about the future together into logical chains. They can add structure to our thought processes. This can help us avoid paralyzing confusion when envisioning Anthropocene tomorrows. But we need not limit ourselves to input/output; we could try doing the same thing with a variety of other simple two-part logical patterns. We could try *"Even if* X happens, *then* Y can save us" patterns. Or we could try *"Either* X could happen, *or* Y could happen" patterns. When that gets boring, we can try cause/effect patterns: "X could *cause* Y, which would have the *effect* of Z," and so on. Here's an example of how we can use "if/then" patterns to speculate about future asteroid impacts, nuclear wars, human extinctions, and technology investments. Try running it through your mind and, if possible, continuing with it, building on it, and extending it further yourself:

So, *if* an enormous asteroid someday hurtles toward our planet, *then* we run the risk of human extinction. *If* the asteroid happens to be smaller in size, *then* it may burn up in the atmosphere. *If* the asteroid burns up in the atmosphere, *then* our total extinction will be averted. *If* the asteroid does not, *then* we all might die. *If* humans have nuclear weapons at their disposal, however, *then* perhaps they could someday be launched into space to (hopefully) blow the large asteroid to bits. On the other hand, *if* nuclear weapons remain in human hands, *then* the threat of nuclear-war-induced human extinction may, in fact, exceed the threat of any asteroid-induced extinction. *If* this is true, *then* societies need to invest in developing alternative technologies that they can deploy to deflect incoming asteroids. *If* they do so, *then* what?

ACCEPTING FUTILITY, RECKONING DEEP TIME ANYWAY

This chapter showed how Safety Case experts admitted to and hedged against uncertainties by developing multiple potential models of the future. Each model got calibrated to different levels of optimism versus pessimism and plausibility versus implausibility. When we reckon distant futures ourselves, we ought to begin by admitting that there is always a good chance we will be dead wrong. At the same time, we must also hold onto the belief that we can, over time, gradually improve our abilities to connect ideas about the future together into several, increasingly intricate, forecasts. This means staying optimistic that integrating expertise-driven learning and two-part forecasting exercises into our daily routines can, ultimately, help us progress as deep time reckoners. Going down this road, we may never end up becoming quantitative modeling experts ourselves. Yet the learning-journey itself can train our minds to be more comfortable with tremendous uncertainties. It can make us more sophisticated in thinking in speculative, futurological, imaginative ways. This holds even if attaining full certainty about the future is impossible.

Investing some of our personal time in collecting expert-vetted knowledge can draw us toward greater accuracy, even if we never arrive at full certainty. Incorporating scientific findings into our long-termist intellectual workouts can make our forecasts more robust. When reckoning deep time, it is always possible to take more potentialities about the future into account. Safety Case experts did this across multiple versions, or "iterations," of their models. Safety Case modeling, then, can offer lessons in never giving up. It can inspire us to restlessly learn and strive toward impossible intellectual horizons. The Safety Case experts showed how venturing to undertake deep time learning is more useful for cultivating long-term thinking than never embarking in the first place. For us, too, envisioning multiple flawed-yet-still-somewhat-illuminating future scenarios will always trump the darkness of envisioning zero. This attitude of guarded, self-aware, measured intellectual optimism is far too rare throughout society during the deflation of expertise.

So, scientific inquiry can clarify certain features of tomorrow. Doing imaginative thought experiments can help train our minds to reckon deep time. Yet all paths forward remain precarious. Analysis of potential future worlds tends to reveal not only previously unknown certainties but also previously unidentified uncertainties. Deep time reckoning simultaneously generates both more certainty (more knowledge) and more uncertainty (more knowledge of one's own ignorance). Keeping this in mind, we must, like Risto's ants, muster the pluck to pursue deep time learning anyway. As nature writer Robert Macfarlane has said:

> What does our behaviour matter, when *Homo sapiens* will have disappeared from Earth in the blink of a geological eye? Viewed from the perspective of a desert or an ocean, human morality looks absurd—crushed to irrelevance. ... We should resist such inertial thinking; indeed we should urge its opposite—deep time can be a means not of escaping our troubled present, but rather of re-imagining it.[22]

DEEP TIME'S DEEP HUMANITY

This chapter showed how radically complex patterns of thought can emerge from much simpler ones. It showed how the building-block concepts used to reckon deep time can be made up of very familiar patterns alive in countless areas of our everyday lives. The input/output distinction was just one example. This means that even the most spectacular, mind-bending, or novel attempts to envision far futures are, in certain ways, continuous with basic patterns shared by billions of other people across the world. They are just another version of basic patterning processes that persist across human generations. Reflecting on this can instill in us an appreciation for the deep humanity of deep time reckoning projects. This means appreciating how a deep time forecast, at its core, retains a "human, all too human" character, despite the alien unknowability inevitable to far future worlds. Reflecting on this can foster in us an enduring fascination with the potentials and limitations of the human intellect in the face of deep time's complexities. If this fascination were to start trending more widely, it could help publics warm up to the expert ethos, which could help counter the deflation of expertise.

So, when we encounter, say, a multicenturial climate change model, we should challenge ourselves to ask questions like these: In what ways is this spectacularly complex expert forecast similar to, or continuous with, the ordinary patterns of our everyday lives and routines? In what ways is it different? Maybe you are lying in bed at night wondering what will happen to Earth when the Sun burns out five billion years from now. Ask yourself: How can I can scour my own familiar patterns of thinking and organizing my life to find better routes into grasping far futures like this? What limitations will I run into while doing so? What concepts, like input/output, do I share with highly trained experts? Which ones do we not share? When should I defer to experts? At what points do experts' forecasting capabilities break down? If I am an expert on something, where does my expertise end and others' expertise begin?

THE LONGEVITY OF BASIC PATTERNS

Many of the simple patterns that we make and follow each day have been used by people for millennia. They are long-lived. Take, for example, the relation of part and whole. This chapter laid out how Safety Case experts saw the Groundwater Flow model as a part of (or input into) the Radionuclide Transport model as a whole, which was then seen as a part of (or input into) the Biosphere Assessment model as a whole. These part/whole patterns were just as central to Safety Case modeling as input/output patterns were. They are also common in many areas of life; they are not unique to expert thinking. They can just as easily be seen among car mechanics, artists, chefs, construction workers, medical doctors, and countless others. My lungs, for instance, are but one part of my whole body. A candy bar is but one part of a vending machine's whole selection of items. A coal power plant is but one part of a whole energy grid. My keyboard, mouse, printer, and monitor are but parts of my whole personal computer. And so on.

Simple part/whole patterns, like input/output patterns, have had longevity. They have quietly guided societies for millennia. They predate not only the Atomic Age, but also both Finland and the United States. As far back as the ancient Greeks, philosophers reflected on the nature of part/whole relationships. They founded a subdiscipline of philosophy today called "mereology," from the Greek word *meros*, meaning "part." The part/whole pattern's longevity across millennia is testament to its centrality to the human experience. It has had long-term impacts on how humans organize their worlds. Reflecting on this can help us position our own thinking habits in wider timescales. It can help us realize that—even when we think thoughts that, on the surface, seem fully about the present or future—we often do so with a foundation of ancient conceptual distinctions that have endured across millennia.[23]

So, we could, for an intellectual exercise, ask ourselves questions like these: Is the use of input/output or part/whole patterns for Safety Case modeling just one momentary blip in a much longer human history of using input/output and part/whole patterns in general? Will these simple two-part patterns live on for millennia even if today's nuclear technologies are totally forgotten? Or will Anthropocene planetary destruction kill us all off first? How will distant future societies use input/output and part/whole patterns? Will these societies ever come to exist in the first place? Will certain nuclear wastes, radioactive for hundreds of thousands or millions of years, outlive us and our input/output and part/whole patterns alike? How can we better grasp the long-term histories of the linguistic expressions and logical reasoning patterns that organize our views of tomorrow? If more people were to ask themselves questions like these, it could help popularize expert-inspired, self-reflective acts of future-gazing. Such popularization could help reverse the deflation of expertise.

DEEP TIME'S DOUBLENESS

We have seen how far future forecasts are often entangled with both the present moment and the deeper history of human thought. To appreciate this is to appreciate the two-faced character of deep time. There is, after all, a doubleness to deep time: human worlds can be inside of deep time and deep time can, simultaneously, also be inside of human worlds. Let's first consider the ways that humanity is inside deep time. We are the temporary outcomes of billions of years of biological evolution. We dwell in ecosystems that emerged from around 4.5 billion years of Earth's geological-climatological history. We are but a momentary episode in the story of a universe that began with the Big Bang almost 14 billion years ago. In these regards, deep history is the stage on which our human dramas play out. It is the setting, context, or backdrop in which our everyday lives take place. We live inside of it.

At the same time, deep time is also inside of our human worlds. For me as an anthropologist, deep time was merely a series of artifacts at my field site. It was a collection of mostly human-made things that I found and observed while doing in-person fieldwork. Sometimes deep time was a pile of Posiva's teal-and-yellow paper reports or a digital folder of PDFs. At other times it was a geologic timetable on my informants' office walls. Sometimes deep time was a topic of discussion in Safety Case experts' hallway conversations. At other times it was a highly technical performance assessment model—or a base, variant, or disturbance scenario portraying a future. Sometimes deep time was an input/output relationship. At other times it was an old rock on the ground. Still other times it was a grabby pop science trope evoking imageries of horror and sublime unknowability—just as it was in Madsen's *Into Eternity* documentary.

Yet deep time was always part of our "phenomenal world impacting on people at the level of experience."[24] Safety Case models were mundane, concrete, real-world products of human analysis. They were electronic or paper "artifacts of modern knowledge"[25] circulating within human institutions. They emerged from a webbed-together network of experts, technologies, communication pathways, reports, formalities, routines, schedules, deadlines, ideas, infrastructures, ecosystems, administrative staffs, customs, norms, emotions, anticipations, and traces of the past.

Sometimes their deep time got entangled with the short-term futures of Posiva's project funding conditions, or the inner workings of interpersonal office politics (as we will see in chapter 4). At other times it got molded by Posiva's plans to maintain stable successions of personnel recruitments and retirements, information transfer and training, and new iterations of the Safety Case until repository decommissioning around 2120.

All in all, then, the deep time I observed during fieldwork was thoroughly embedded in human worlds. Yet these human worlds were, in another sense, thoroughly embedded in deep time. They were but momentary episodes of our species', planet's, and universe's deeper histories. Challenging our intellects to toggle back and forth between these two senses of deep time—first as a timeframe that humans inhabit, and second as an artifact that humans create—can help us view the Earth's radical long-term through an anthropological lens. This can serve as a foundation for developing the more multidimensional, multitimescale form of deep time reckoning that the next chapter argues becomes imperative during the Anthropocene and the deflation of expertise.

3

HOW TO ZOOM IN AND OUT ON DEEP TIME FROM DIFFERENT ANGLES

2013 CE, field notes: Late one night in October, I woke up from a restless dream. It was centuries ago; I was part of a military platoon. We were preparing to raid a nondescript foreign enemy. The enemy was based at an encampment in a forest outside Helsinki. As we marched, the landscape was flat. It was covered mostly by pine trees of uniform height, but also spruce and birch trees. It was summertime, around 10 p.m., and the sun was beginning to set. We knew the forest was home to hare, lynx, deer, bear, wolves, and flying squirrels; we saw none. Our noisy troop must have scared them away. As we drew closer, moving quietly, we peered through the shrubs at our foes. Then they spotted us. Our enemies began shouting at us, and we at them. Weapons in hand, we exchanged threats in two different, mutually incomprehensible tongues. Then, out of nowhere, my perspective switched. Suddenly, I was seeing the world from my enemy's point of view. I was looking at myself in the third person: watching me stand there, waving my fist in the air, yelling aggressively in Finnish, clutching my gun. I felt ridiculous for embarking on the raid in the first place. Shifting my perspective, and changing the angle from which I viewed the situation, had altered my sense of mission. I wanted to resolve the conflict peacefully. But then, just as suddenly, my perspective shifted again. I woke up. It was all a dream. From the perspective of my bed, the whole situation felt inconsequential. It had seemed so gravely important just moments ago. Lying there, I began thinking about how a simple change in viewpoint can reorient how we approach the challenges we face.

ANGLES AND SCALES

I once asked a Safety Case expert named Laura, "What is the most challenging thing about your work?" She replied:

It's the one million years. That is so mind-boggling. It is much longer than anything we have known in recorded history. And then think that we have to build canisters that are supposed to last hundreds of thousands of years. Nothing has lasted so long, nothing. We have stuff built by man that is two, three, four thousand years old. There are [rock] paintings in Finland, but how old are these [rock] paintings? A few thousand years. That's nothing. It is this timeframe that is really hard to understand for us and only geology can help us with this type of timeframe because we get into the ore formations, like copper ore formation and geologic-like movements, the continents coming together. ... Sometimes I think I would like to be immortal just to see what happens in one hundred thousand years to the canisters and then, hopefully, be like, "Ah, you see, I told you so!"

Laura had grown up in Italy, getting her PhD in chemistry in France. She had worked for a government funded agency there that focused on defense, energy, and security research. After that, she had worked in the United States for a large national science policy research organization. She now lived in Finland with her American husband, who was also a scientist with a PhD. He did laboratory research on bentonite clay for Finland's repository project. Laura's backstory was testament to the sort of highly specialized, globally mobile, policy-aware, and transnationally connected expertise optimal not only for managing nuclear waste, but also for reckoning deep time. Recalling the factors that led her to settle in Helsinki, Laura told me how, in Finland, workplace culture allowed her to take her kids to the dentist without first having to beg her bosses for permission to do so. Her kids could play safely in the streets without her worrying about crime.

When I met Laura, she was helping develop a Safety Case report that gave an overview of how the portfolio's many models, datasets, and reports wove together. It was titled *Models and Data*. Later Laura would become one of Posiva's lead research managers. To put Posiva's far future forecasts into perspective, she reached back into Finland's human past. She reflected on its ancient rock paintings and how continents came together and separated over millions upon millions of years. Doing so helped her appreciate not only her work's immense time spans but also the Safety Case's immense ambition as a human project. Like Laura, other Safety Case experts were also fascinated by how happenings of the past had developed historically into the moments in time that Finns now inhabit. They too saw the past as a font of learning. To quote a Finnish geologist, reflecting on his youth:

We used to dig up old German helmets. There were some friends of mine, and we were all are interested in history. And there are some famous fighting places from the Finnish-German War in Lapland: Lapland's War. There was one spot where there were heavy fights. The Finnish were coming from the south, and the Germans had their headquarters or their frontline there. It was a small straight. There were lakes and there and a small escarpment ... maybe two hundred people probably died there. And you can still find the remnants of it. We found one place where it was this evacuation place for wounded Austrian German soldiers. We looked through old maps. I was like twenty or something at the time. And when we had a metal detector with us, we found some even medals and some kinds of historical objects. ... Another time we found this trash site for German soldiers. It had cans and even some blades and some mugs. Back then, the Germans always put some serial number or some eagle on every object, every screw, every small thing. And we found dozens. We found tin cans from Croatia, from Hungary, and wine from France because the labels were still there. So actually, you could see that Germany's Army was really well, you know, how to say it, *equipped* ... because of the foodstuff from all over Europe, conquered lands ... they have been drinking French champagne here.

For this geologist, human histories were sources of intellectual intrigue. He told me about how digging up military trash as a youth motivated him to help bury nuclear waste in the ground at work as an adult. For other informants, however, human histories' complex contingencies were reasons for uncertainty and dread. They reflected on the Safety Case's position within the Earth's longer-term timescales, expressing concern for the Olkiluoto repository's fate. One informant told me:

One thing that cannot be handled well over these distant future timescales is the human. What kind of society will we have? Will we have humans here after ten thousand years? Wars? These kinds of scenarios have not been taken into account very carefully. I am afraid of them. We know we can understand the physical world as engineers, but the human is the part we cannot understand. What will they do with nuclear waste? You don't know what will happen to humankind or Finnish civilization.

Deep time reckoners like these are skilled at *toggling back and forth* between visions of human, ecological, and geological pasts (near and distant) and human, ecological, and geological futures (near and distant). As in many other lines of thinking during the Anthropocene, distinctions made between these timescales tend to blur into one another. Learning to better perform these intellectual gymnastics in *zooming in and out* across time is, my fieldwork taught me, a key Anthropocene talent. Pursuing thought experiments in learning this talent from the world's most longsighted experts can help us counter the deflation of expertise, especially if this practice someday comes to gain public support.

To this end, this chapter will take three learning-journeys across time. First, it will *zoom way out* by approaching the entire Safety Case project as one momentary blip in the deeper history of Finland. Doing so will provide a bigger-picture basis for appreciating how the Safety Case itself was a product of history. It will show Posiva's repository work to be a brief episode that emerged from a dizzying array of past causes and effects that occurred over decades, centuries, millennia, hundreds of millennia, and beyond. After that, the chapter will *zoom way in* on the Safety Case's everyday workplace project timescales as they appeared when I studied them anthropologically from 2012 to 2014. Doing so will show how the Safety Case organized a complex array of expert perspectives together into a workable collaboration over the project's months and years. Third, it will *take a step back* from Safety Case experts' professional time and explore what they did in their personal leisure time, to show how informants recharged their minds after work. This recharging helped them avoid occupational burnout and sustain the energy necessary for persisting with knowledge-intensive Safety Case work, day in and day out.

Learning to hop more nimbly around these timescales can inspire a more refined *multiscale, multiangle,* or *multiperspective sensibility*. This kind of multidimensional thinking must, I suggest, be cultivated during the Anthropocene. For the purposes of this case study, this means learning to swing from Finland's deep time to the Safety Case project's months and years to my informants' day-to-day lives. As in chapters 1 and 2, I will close this chapter with five reckonings. However, here, and in the next chapter too, the reckonings have a different target. They will focus less on how individuals can do introspective thought experiments and become shrewder deep time reckoners themselves. Instead, they challenge us to ask: How can today's societies better support the highly trained, prolific deep time reckoning experts already in our midst? Can existing deep time reckoning experts, the organizations that employ them, and the infrastructures, colleagues, and administrators working alongside them adopt new policies, programs, or working culture norms to foster long-termism? How can organizations with long-term impacts—from plastics manufacturers to chemicals companies to the fossil fuel industry to the financial services sector—reform their attitudes toward time? Future changes in lay and expert cultures alike are needed to avert planetary collapse. That is why deep time reckoning requires greater institutional support. With this in mind, let's stretch out our minds by traveling back in time again, taking a learning-journey to an alien Earth as it appeared over seventy millennia ago.

ZOOMING OUT: FINLANDS IN FORMATION

Our story of Finland's long-term history can begin around 70,000 BCE, with the expansion of the enormous glacial ice sheet that covered it throughout the previous Ice Age. At its peak, the ice was three kilometers thick. Dubbed the Weichselian glaciation, the ice sheet once covered most of what are now called Europe's Nordic countries. About 20,000 years ago, it began its retreat. After years of melting, Finland's glacial period ended around 9000 BCE. With an immense weight off its shoulders, the landmass we today call Finland began rising. Its coastlines extended outward, gaining ground. Settlers moved to Finland shortly after the ice sheet retreated, most likely from the south and southeast. The climate was cold and dry. Birch forests displaced tundra. Coastal areas near what is today called the Gulf of Finland emerged from the ice around 10,000 BCE. Glacial melting hastened rapidly a thousand years later, leaving behind the Salpausselkä ridge in Southern Finland. This region has seen human habitation for more than 10,000 years. By 7000 BCE, the ice sheet was gone; deciduous and pine trees sprouted up in droves.

Seminomadic communities hunted, fished, picked berries, and foraged for plants. Inhabitants used elk bone to make spearheads and ice picks. They used elk antler to make axes and combs, elk hooves to make cups, tendons for string, bladder for water conditions, brains for tanning, and eyes as a binder for painting.

By 6500 BCE, most of Finland was covered with pine forests. Pottery found in Finland dates back to 5300 BCE. Around 4000 BCE, parts of Finland were as mild, wet, and warm as present-day Central Europe. Clay pottery-making practices were adopted from the east. About 500 years later, some inhabitants began decorating and embroidering their clay-ware using combs. Today's archaeologists call these the "Comb Ceramic cultures" or "Pit–Comb Ware cultures." Human and animal populations thrived. Luxury adornments made out of amber were imported from the Baltic Sea's southeastern shores. Furs from Northern Finland made their way to the South via trade routes. Many humans' tools were adorned with figures of elk, waterfowl, and bear heads. Over the years, various inhabitants left behind rock paintings, which captured the Safety Case chemist Laura's imagination in 2013 CE. The rock paintings depict images of snakes, people, moose, boats, fish, birds, and more.

Around 2500 BCE, Finland's climate grew drier and cooler. Many of its deciduous oak, alder, and birch trees died; spruce trees sprouted up and spread. A new group of Indo-European peoples arrived in Southeastern Finland. They brought new languages and animal husbandry practices. They made battle axes shaped like boats and adorned their pottery with cord-like ornamentation. Finland's Bronze Age, beginning sometime after 1500 BCE, drew Germanic influences into the region through Scandinavian immigration and coastal contacts in the west. "Nordic Bronze Culture" influenced coastal regions of Finland. Finland's Iron Age (~500 BCE to ~1300 CE) saw increased contact with the Baltic regions, which today comprise the countries of Estonia, Lithuania, and Latvia. Finland's climate became colder and wetter around the dawn of the Common Era.

At the end of the Iron Age, one could discern a few distinct language dialect groups across Finland: *Hämälaiset* (Tavastia), *Varsinaissuomalaiset* (Finland Proper), *Karjalaiset* (Karelia), and *Saami* (Lapland). During pre-Christian times, Finns and Vikings were known to one another through plundering and trade. Some archaeologists have noted evidence of

Swedish settlement in Finland's southwest. Christianity gained a foothold in the tenth and eleventh centuries CE. Finland's Catholic medieval period began around the twelfth century and continued until the Reformation in the 1520s. Practices of writing down events came to Finland only after the rise of Christianity. By the thirteenth century, Finland was a key battleground between Orthodox Russia and Catholic Sweden.

Most regions of what is today called Finland were part of the Swedish Empire from the thirteenth century until 1809. Lutheran clergyman Mikael Agricola established a comprehensive writing system for the Finnish language in the 1500s. Parish population registers were established in the 1600s, keeping records of Finns' births, marriages, and deaths. From 1695 to 1697, Finland suffered a massive famine that killed about one-third of its population in just two years. From 1866 to 1868, Finland faced *suuret nälkävuodet* ("the great hunger years")—a famine that killed almost 9 percent of its population in three years. In 1809, most Finnish-speaking regions were ceded to the Russian Empire, and the autonomous Grand Duchy of Finland was established.

Finland declared independence from Russia in 1917. Soon after, a civil war flared up. The anticommunist White Guard defeated the socialist Reds. From 1939 to 1940, Finland defended itself from Soviet invasion in the Winter War. From 1941 to 1944, Finland fought the USSR as a cobelligerent aligned with the Axis powers. In 1944 and 1945, after signing an armistice agreement with the USSR and the United Kingdom, Finland fought the Lapland War. This was an effort to expel Nazi German forces from Northern Finland. The helmets and wine bottles that the German occupation left behind would, decades later, capture the imagination of my geologist informant when he unearthed them in his youth. By the end of these conflicts, Finland was forced to cede, as reparations to the Soviet Union, not only its Petsamo region, but also parts of Karelia and Salla, plus four islands as well.

Since then, Finland's economy developed to boast one of the world's highest per capita GDPs. Finland's welfare state expanded significantly between 1970 and 1990. Two nuclear reactors, in the municipality of Loviisa on Finland's southern coast, were turned on in 1977 and 1980. Some industry insiders playfully nicknamed Fortum's Loviisa power plant "Eastinghouse" because its Soviet-designed reactors were fitted with

Western safety systems and controls made by the Westinghouse and Siemens corporations. Two other nuclear reactors were located right next to Posiva's repository on Olkiluoto Island. They began commercial operation in 1979 and 1982. Finland's research reactor at Helsinki University of Technology (now Aalto University) was originally purchased from the United States. Located in Espoo right outside Helsinki, it was turned on in 1962. Responsibilities for it were transferred to VTT Technical Research Centre of Finland in 1971. During my fieldwork, Finns still had memories of the 1986 Chernobyl nuclear disaster's fallout raining down on their country's soil. Cesium from Chernobyl was still detectable in Finland's mushrooms, elk, and reindeer meat.

After the collapse of the Soviet Union, an economic recession welled up in Finland. This turmoil was mostly resolved by the rise of Finnish mobile phone and IT companies, including Nokia. The nuclear waste management company Posiva was established in 1995. Finland joined the European Union in 1996, and adopted the euro in 2002. Olkiluoto was formally chosen as Posiva's repository site in 2000. Excavation of Posiva's Onkalo underground laboratory began in 2004. After years of twenty-first-century decline, Nokia was sold in 2013 to Microsoft for €5.44b. During my fieldwork, many spoke of the retraction of Finland's traditional pulp, forestry, and paper-processing industries. They spoke of Finland's struggling mobile phone businesses. Many wondered, with guarded anxiety, what the country's next "national project" would be.

Meanwhile, Finland's new Olkiluoto 3 reactor was still not yet operational. It saw serious cost overruns. It was originally slated to go online in 2009. The Finnish power company purchasing the reactor, TVO—and the French-led Areva-Siemens consortium that signed on to design and build it—were suing one another for billions of euros in compensation. TVO's plan for a fourth reactor in Olkiluoto was abandoned by Finland's government in 2014. The plan may be revisited someday. At the same time, Fennovoima was seeking an entirely new power plant called Hanhikivi 1. Fennovoima signed on Russia's Rosatom as its technology supplier. Established in 2007, Fennovoima was Finland's newest nuclear power company. When I was conducting fieldwork, TVO, Fortum, and Posiva sought to prevent competitor Fennovoima from having access to burying its spent fuel in their Olkiluoto repository. As I write in 2020, the Olkiluoto 3 reactor remains a costly work in progress. Fennovoima's new reactor is expected to see further delays. It is now scheduled to begin operations in 2024.

Learning more about longer-term histories like these can help us, as it did for informants like Laura, to put both the Safety Case project and its far future forecasts into a broader perspective. After all, every event I just covered occurred only within the last 72,000 years. Finland's commercial nuclear reactors were built only in the past half century. Finland's first repository safety analysis was published only in 1985. In short, all of this occurred within timescales that were but tiny fractions of those forecasted by, say, Posiva's Radionuclide Transport model. The transport model looked hundreds of thousands of years into the future.

These historical events comprised even smaller fractions of the timescales covered in the Safety Case scenario *The Evolution of the Repository System beyond a Million Years in the Future.* They comprised a remarkably small fraction of the time since the Lappajärvi meteor slammed into what we now call Finland about 73 million years ago. From the long-termist vantage point of someone who has achieved immortality—to use Laura's example, quoted above—the Safety Case, and Finland itself, have been instantaneous episodes in Earth's deeper history. Comparing the brevity of human pasts with the Safety Case experts' ambitions to reckon far futures can, informants noted, provide a "mindboggling" encounter with deep time's breadth. As philosopher Edmund Burke put it back in 1757: "Greatness of dimension, vastness of extent, or quality, has the most striking effect."[1]

However, zooming way out to view the Safety Case project as a mere instant in history can also help us appreciate the countless past events that first had to elapse in order for the specific cast of characters I met in Finland to have been born, encultured, and educated into the expert people they ultimately became. After all, if we were to hop into a time machine and change the course of events that led up to their births, my fieldwork informants may have never come to exist in the first place. If Finland's multimillennial human migration and famine histories had occurred differently, the composition of the country's population may have turned out so different that their parents may never have met (or existed). This could also be the case if Finland's Civil War, World War II, or the Cold War had played out differently. If Finland had spent the Cold War as part of the Soviet Union, the twenty-first-century economic conditions and cultural mentalities I observed there may have evolved into something quite different. This ethnography would, as a consequence, read very differently. If the Finnish government's immigration policy histories had been different, Laura may never have been allowed to settle there. If twentieth-century globalization trends had been less powerful, Laura may have never received her uniquely tailored international training. If Finland had not seen a post–World War II economic upswing, the country may not have been able to afford nuclear reactors. Having no spent fuel to manage, my informants would likely be working in other fields, living rather different lives. That is, in the unlikely event that they would exist in any recognizable form in the first place.

In other words, the Safety Case experts' lives were, in these ways, products of histories. They were contingent on, or culminations of, endlessly complex chains of causes and effects that wound far back through time. This included deep histories. For instance, if Finland's glacial ice sheet had never melted, the country may have stayed uninhabited altogether. If Finland's bedrock had never formed the way it did, Posiva may have chosen a different repository type. The crystalline granite bedrock's hardness and durability was, after all, central to Sweden's and Finland's KBS-3 design's functioning. Zooming out and viewing the Safety Case from this longer-term standpoint can help us appreciate how an unfathomably huge series of past events were always alive in the Safety Case project. It can help us reflect on how my informants' collaborations emerged out of distinct historical circumstances that brought a specific group of people, material conditions, technologies, expertise, and government apparatuses into being.

Viewing the present from a longer-term geological or historical perspective can help us take a step back. It can, in geologist Marcia Bjornerud's terms, teach us "timefulness." That is, it can make us more "mindful that this world contains so many earlier ones, all still with us in some way—in the rocks beneath our feet, in the air we breathe, in every cell of our body."[2] With that in mind, let's now toggle way back in the other direction in time, as my Safety Case informants often did. Let's extend an adventurous spirit of deep time learning to the very short-term forces that held the Safety Case collaboration together. These short-terms spanned only days, weeks, months, and a few years. They are the mundane time horizons that structured Posiva's deep time reckoning projects, as they geared up for submitting their construction license application to regulator STUK.

ZOOMING IN: FORESTS AND NAZCA LINES

Chapter 2 described how, when Safety Case experts collaborated together, no single expert was able to simultaneously comprehend the details of all of Posiva's individual reports plus how these many reports linked together to form a whole portfolio. As one put it, "No one person can describe how the whole system works based on what we know anymore." Given the Safety Case's tremendous growth over time, it had to be reckoned by a group.

Pondering this, I asked a geologist named Taimi how she personally saw her work feeding into the Safety Case project.[3] She, like Laura, answered me by reaching back into human history. Taimi compared the Safety Case collaboration to the building of Peru's Nazca Lines geoglyphs between 400 and 600 CE. Taking a zoomed-in view from the ground, she explained, the Nazca Lines merely look like long walls or arbitrary lines of stones. Taking a zoomed-out view from a helicopter, however, the Nazca Lines take form as images of hummingbirds, monkeys, lizards, and sharks. To create the Nazca Lines or to create the Safety Case, Taimi reasoned, each collaborator must understand how the smallest parts of one's own work will be *scaled up* to feed into the project's bigger picture:

As for the Nazca Lines, flying above you can see that there are patterns. Nobody knows how they were made, but you can see their broader logic only from above on an airplane. They're unremarkable when you view them from the ground. To figure out what they represent on the ground, you would have to case the things and sketch it and make measurements. ... Viewing the Nazca Lines from an airplane, you are not lost in the rainforest trying to understand the whole by tracing it out from a point or single report, a single part of the larger fabric.

Taimi's doctorate was in geophysics and engineering geology. But she shared an interest in archaeology with her daughter, who was studying it at university. Her Nazca Lines analogy gestured to how each Safety Case expert's highly specialized work (just one part of just one Nazca Line) contributed to a larger group achievement (the powerful images visible to those who view the Nazca Lines from on high). Taimi sometimes discussed this with her colleagues. Her aim was to get each Safety Case expert feeling as though he or she was building something larger than him- or herself: something greater than the sum of its parts. This challenged individual experts to zoom out from their everyday perspectives. It nudged them toward taking a step back from their own work. It dared them to try to see their own work as it would appear from a vantage point outside of themselves—viewing the totality of the Safety Case collaboration all at once, from on high.

The trick, for Taimi, was to train one's intellect to zoom in and out between the ground-level details of one's own work and its airplane-level overview. This could get one appreciating both (a) the inevitable incompleteness of any single expert's perspective and (b) the great mutual reliance necessary among experts to reckon deep time. Adopting this sensibility was, in her view, key to achieving effective teamwork relationships. This became especially important in a project that could be understood only when many different experts with many different perspectives were viewing the Safety Case from many different standpoints all at once.

I also brought up these issues of project coordination and scale with Laura. She replied with an analogy of her own, and compared the Safety Case collaboration to a forest. Like Taimi's Nazca Lines analogy, this analogy called on experts to do an intellectual exercise of zooming in and out on one's own perspectives. The aim was to achieve a more holistic, or bigger-picture, understanding:

In the Safety Case, you have the forest and the trees. The forest is the big overview; the trees are the details. ... When you say, "You're looking at the trees, you're not looking at the forest," we mean "Look at the big picture, don't get bogged down focusing on the details." And that's what we do. We try to look at the big picture. Try to keep in mind the whole ecosystem, not the individual details that can go very deep down to the roots. That's where the modelers work, to make sure that everything works. If you have only the top without knowing a little about the roots you have just this big green mass. And then if you have only the roots, you only see grass and brown stuff and you don't have the whole gamut. ... Seeing both comes with experience and time. ... You cannot isolate one branch from the rest of the tree. It has to be organic. We need the food from the roots. And the roots need the treetops for light and life—for money to do their research.

Laura's forest analogy emphasized how it was the job of some experts and managers to take a zoomed-out view of the portfolio's "treetops." This meant ensuring that more broadly scaled reports properly grasped how the Safety Case's many models, datasets, scenarios, and engineering designs connected together. These metalevel reports included the *Synthesis* and *Models and Data* reports discussed in chapter 2. Other experts, meanwhile, were to take a zoomed-in view of the portfolio's "roots." This meant grasping the details of how, say, highly specific datasets (e.g., data about groundwater chemistry or fish populations in Western Finland) got fed into Safety Case models. Still other experts were to zoom to a middle level of the portfolio's "branches." This referred to the reports and models that, to use chapter 2's terms, took in other models' outputs as inputs of their own, but still were not at the end of the Safety Case model chain. These "branches" models still produced outputs that fed into other, more-encompassing metalevel models closer to the treetops.

For the Safety Case forest to maintain its aura of comprehensibility over the months and years, it needed to be viewed by dozens upon dozens of living professionals from many different perspectives, scales, and levels. Laura sought to help build up this multiperspective spirit of teamwork. Her forest analogy drew her colleagues toward seeing themselves as inhabiting a network of interconnected workplace roles. They were to view themselves as responsible people positioned within a larger division or hierarchy of responsibilities. The challenge, for each individual expert, was to better grasp how their unique perspectives were positioned in relation to others' unique perspectives and vice versa.

Like Taimi's Nazca Lines analogy, Laura's forest analogy highlighted the impossibility of any one human mind simultaneously comprehending the full intricacies of each and every part of the Safety Case, and also how those parts wove together to form a whole portfolio. She nudged her colleagues toward reflecting on how their projects were nestled in specific niches within a much larger collaborative ecosystem, and encouraged them to avoid getting bogged down in details that may be insignificant to the Safety Case's bigger picture. For her, one should routinely attempt to envision oneself zoomed out from one's own trees (one's individual projects) by considering them in light of a larger forest (the Safety Case collaboration as a whole). One should try to imagine how one's work would appear when viewed from other Safety Case colleagues' perspectives within a larger forest of work. This means trying to grasp other perspectives zoomed in and zoomed out on various positions in the project's roots, branches, or treetops. Doing this could, for Laura, help experts reflect regularly on the limits of their own expertise, and get them thinking about how other experts' expertise filled the gaps in their own knowledge.

Laura's and Taimi's analogies had something common: they both advocated routine efforts to shift the angle and/or scale of one's perspective. This meant intentionally altering the perspective—and level of generality or specificity—from which one approaches a problem. Doing these thought exercises helped the Safety Case experts attain a more robust sense of how to pursue their grandest ambition: to reckon far future Finlands with greater sophistication. The ultimate goal of these shifts in perspective (that is, to more wisely organize an expert collaboration in short-term timescales) was different from that of the previous section's shifts in perspective (that is, to take a step back and gain a deeper perspective on the Safety Case project by viewing it in deep timescales), but the intellectual route there was similar. Both were about trying to view a set of present-day challenges afresh by zooming in and out on them from many different vantage points. Both were about embarking on holistic, adventurous learning—considering and valuing all sorts of different perspectives.

As the next section will show, a related set of perspective-shifting practices were at play in how certain informants *took steps back from*, or zoomed out from, their daily professional labors. Doing so, they refreshed themselves and their intellects during their personal leisure time. This zooming out helped them *achieve distance* from their work before they "circled back"[4] to it again the next day and saw it afresh. With that in mind, let's now explore how Safety Case experts saw time outside the workplace as a space for rejuvenating themselves. This supported their aspirations as deep time reckoners, enhancing their ambitiousness, inner calm, and alertness when they returned to their highly technical Safety Case work the next day.

COUNTERPOINTS: TAKING A STEP BACK

The Safety Case's visions of far future worlds emerged from a host of unique expert perspectives, by bringing together a host of specialized, fallible, distinctive people to collaborate together in a particular place at a particular time. Some experts, as one informant put it, had more "ancient Greek" mentalities (more interested in hard logical relations, data, calculation, and syllogisms). Others had more "ancient Mesopotamian" mentalities (more interested in soft intuitions, broader phenomena, and general patterns of relationship). Some experts were better at skating perfect figure eights (executing procedures perfectly or performing research with rigorous exactitude). Others were better at improvising like a jazz musician (inventively troubleshooting in ad hoc or on-the-spot ways). Some higherups could be good managers (organizing people, projects, knowledge, and things efficiently) but were poor leaders (failing to inspire a team's morale or self-motivations). Some experts had a calm, sober, disciplined interior that enabled them to keep a cool head when things went wrong. Others became panicky and flailed under pressure. Some tended to be eccentric, jumpy, or nonconformist but became focused, serious, and dependable when challenges arose. Some were geologists, others mathematicians, others engineers, others chemists, others physicists, and so on. Different types of scientists understood and troubleshot the Safety Case in different ways. It took a village to make far future Finlands appear.

Each citizen in the Safety Case village had his or her own ways of replenishing, or failing to replenish, his or her inner reserves of energy outside work. These activities helped keep them going. One soon-to-retire manager described how he enjoyed cooking, yard work, gardening, and making renovations to his home. He listened to the music of Estonian composer Arvo Pärt. Baroque chamber music helped him achieve a meditative calmness. An engineer who worked on the repository canister design took winter breaks vacationing in the Canary Islands. In the summer months, many informants relaxed in rural Finland at their family's lakeside *kesämökki* ("summer cottage"). A hydrogeologist in her early forties saw horseback riding as an enjoyable counterpoint to her expertise, which helped her clear her mind in the forest. A biologist in his thirties described how long-distance running helped him "empty his head" and made his "thinking" processes "vanish." The biologist saw the sauna not only as a place where he could "think more clearly," but as a "holy place" for Finns: "We do it from the beginning when we are children. ... It's the thing that's constant throughout Finns' lives. ... Sauna is always sauna." A physicist in his sixties started taking group acrylic painting lessons after his mother and daughter bought him art supplies as a gift. He kept a picture of a bear on his wall, looking at it daily to try to get a clearer "mental model" of its details, so he could paint it. The physicist contrasted painting with his scientific work, seeing painting as offering more "freedom," allowing him to "release" his mind from the "limitations" his scientific work imposed.

These many activities seemed, on the surface, to be unrelated to the Safety Case experts' professional projects. They could be athletic, artistic, or musical in character—or even tasks resembling physical labor. But they were all essential to keeping the experts going. They provided my informants with counterpoints to their work: spaces of refuge that enabled them to routinely take breaks from, take steps back from, or achieve distance from their knowledge-intensive and often mentally taxing workplace projects. Some of these activities were refreshing because they had clear endpoints or discrete final products—like planting flowers or cooking a meal. This contrasted with the air of neverendingness that Safety Case work entailed. Other activities, such as painting, were relaxing because the stakes were much lower than those of managing nuclear waste.

The key, for these informants, was to carve out places and times in which overactive minds could take breaks. This meant focusing on non-work-related hobbies that helped their wound-up ruminations subside. The informants I met who did not have (or who simply did not have time for) these counterpoint activities appeared more likely to struggle with occupational burnout, exhaustion, or pessimism, which could dampen their aspirations to reckon deep time. To keep on growing, the Safety Case needed to be continually infused with hard work, intellectual energy, and rigor by its living authors. As chapter 4 will show, the project depended on engaged human physiologies and alert expert minds to persist. Declines in team morale could mean a stagnant collaboration. I once observed a nuclear energy industry insider with a background in the US nuclear navy put it this way:

When you take a look at a nuclear reactor, you realize that we as nuclear workers have a unique responsibility to take care of that core above all else. When you don't protect the core, bad things happen. So, our challenge is to become the conscience, the voice, the person who cares for what happens inside of the reactor core. This is inherent in the nuclear business. But to do this, one must also protect one's personal core, the core of one's body, the core of who one is. All of us have a personal core: it is who we are, what we're made of. This is equally as important as protecting the reactor core. ... Respect the personal core or else bad things happen.

To protect their personal cores, many informants spent leisure time zooming out from their sometimes nerve-wracking technical work. They were not alone. Legal scholar Alan Dershowitz has cast "play" as his key to achieving "tremendous amounts of energy," noting his love for "opera, hiking and museums." Filmmaker David Lynch has described the challenge of overcoming the "heavy weight of negativity," which, to him, is the "enemy" of creativity. Lynch has also embraced quiet meditation practices. Reflecting in 1917 on how "enthusiasm" and "work" combine to "entice" ideas from the mind of the scientist, sociologist Max Weber declared that "ideas occur to us when they please, not when it pleases us." Big ideas, for Weber, were more likely to sneak up on us when walking down the street or "smoking a cigar on the sofa" than when "brooding and searching at our desks." At the same time, a fresh idea would be stopped in its tracks "had we not brooded at our desks and searched for answers with passionate devotion" in the first place.[5]

In this same vein, my informants' leisurely counterpoint activities did not compete with their expert work, and neither did the feelings of personal replenishment that their extracurricular activities delivered. Rather, these elements were essential to the success of the Safety Case project. This may seem counterintuitive. As Pope Francis has lamented, many people today "demean contemplative rest as something unproductive and unnecessary."[6] Yet, when it comes to reckoning deep time, having no time to think can stultify the intellect and drain creativity. This can deprive a Safety Case expert of the dynamism that comes with taking a step back from work, achieving some distance, and then circling back to it the next day. There is so much to be gained from viewing one's ideas in a new light, from a fresh perspective.

SHIFTING SCALES, SHIFTING PERSPECTIVES

This chapter presented what social scientists might call a multiscalar perspective.[7] It put forth an analysis that tacked back and forth between several scales of generality and particularity, approaching the topic from many different angles and levels. It began by viewing the Safety Case from the zoomed-out perspective of Finland's multimillennial human and geological histories. It then zoomed in on how Posiva's collaborations achieved more robust organization across weeks, months, decades, and years. Next, it zoomed in even further on how Safety Case experts maintained their motivations to endure their work's sometimes exhausting day-to-day demands. This could be called a multitemporal analysis: it approached a case study from a variety of different time spans simultaneously.[8] Yet this chapter also explored the value of approaching problems from multiple angles, perspectives, and viewpoints. Taimi's Nazca Lines analogy and Laura's forest analogy helped us understand how such a diverse group of experts came together to form a project greater than the sum of its parts. This required them to work toward viewing the Safety Case from different positions, roles, and disciplines. What emerged was a more dynamic, multidimensional form of deep time reckoning.

The most skilled deep time reckoners were adept at hopping around between different scales of time. They attempted to see their own thinking from others' perspectives. They appreciated how any technical issue can be understood in many different lights. They knew when it was time to achieve distance from their work's hairsplitting details. They knew when to dig deeper into its weeds. They knew how to avoid going too far down any one rabbit hole of thought. They knew when and how to take breaks, too—taking a step back from their knowledge-intensive professions and personal ruminations.

Yet one does not need a PhD, or even a high school diploma, to cultivate these intuitions: anyone can embark on a learning-journey of discerning how any given event, entity, or vision of the future can be viewed in several different ways—depending on the perspective from which one approaches it. Inspired by this, we can begin asking how today's short-sighted organizations, especially those with long-term impacts, can adopt more multitemporal worldviews. We can ask how communities of skilled deep time reckoners can be given more say in today's societies. The big question is how to better orchestrate experts' and citizens' relationships with time, with one another, and with the Earth's future ecosystems. This means ensuring that experts and laypeople alike work toward envisioning future worlds from several different scales, levels, and vantage points simultaneously.

RECKONINGS

MULTITIMESCALE AWARENESS TRAINING

The ways Safety Case experts toggled back and forth between histories and futures near and deep can serve as models for personnel training courses at any corporation, agency, or research institute with long-term impacts. Companies could host workplace retreats, boot camps, or office workshops that cultivate their staff's long-term thinking patterns and multitimescale awareness. These could be taught by geologists, climate scientists, cosmologists, anthropologists, nuclear waste experts, archivists, astronomers, philosophers, or other kinds of deep time reckoning experts. Casting long-sighted physical scientists, social scientists, and humanities scholars as professional mentors for lifelong learning can do more than just raise awareness about Anthropocene perils; it can also help empower the virtues of careful and rigorous expertise. This empowerment, if accepted by the wider public, can help counter the deflation of expertise.

For example, teams from a fossil fuel company could learn all about the deep geological history of West Texas's Permian Basin oil field. They could learn about the long-term human migration and settlement patterns that ended up peopling the oil industry's executive teams and labor forces. Participants could be challenged to reflect on, say, how their organization's short-term timescales of quarterly earnings, executive turnover, and financial transactions relate to, compare with, or have been affected by these deeper histories. They could then envision the fossil fuel industry's longer-term consequences. These range from carbon emissions causing climate change, to the challenge of Earth's limited fossil resources, to how resource extraction can affect local ecosystems. They could cover positive impacts, too: they could reflect on how their organization's ongoing job creation stimulates the economy or how injecting energy into the grid can raise standards of living—extending the timescales of human lifetimes.

Workshop attendees could also be introduced to social scientific perspectives on multitemporality, including sociologist Barbara Adam's books about timescapes. Her work explores how different paces, scales, or "scapes" of time—from democratic electoral schedules, to the tempo of capitalist commodity production, to communication and transportation speeds—converge to stoke environmental problems.[9] To help achieve critical distance from their everyday office lives, participants could then learn about anthropologist E. E. Evans-Pritchard's classic study of time-reckoning practices among Nuer communities in South Sudan.[10] Evans-Pritchard made a distinction between Nuer oecological time (concepts of time derived from local environmental circumstances, like scheduling one's day around the "cattle clock" of milking cows, taking them to the meadow, putting them in a stable, and so on) and Nuer structural time (concepts of time derived from people's relationships with other people, lineages, clans, political groupings, statuses and ages, and so on). Present-day anthropological case studies, such as this chapter's multitemporal analysis, could be discussed as well.

The goal of these exercises would be to nudge employees at all levels and positions toward approaching problems from multiple timescales, which would help the workforce think in spans more in sync with Anthropocene challenges. Enriching one's time consciousness can be eye-opening for any employee. This includes high-level executives, accountants, administrative staff, communications professionals, and others as well. It can remind anyone how, during the deflation of expertise, we cannot ever allow ourselves to stop learning—even after our formal educations are complete. If this sounds unrealistic, just think of how unrealistic diversity training, sexual harassment training, or management training must have seemed to corporate leaders just a few decades ago. Yet these office programs are now commonplace. In any case, even if, say, today's oil companies do not have any interest in implementing multitimescale awareness training programs on their own, voters can still push for laws requiring them to do so.

DAYS, DECADES, CENTURIES, AND MULTIMILLENNIAL DIVISIONS

During the Anthropocene, many private companies have long-term impacts on the planet. These range from plastics manufacturers to fossil fuel extractors to those in the chemicals industry to the financial sector. Yet most companies are remarkably short-lived. In the late 1990s, Arie de Geus of Royal Dutch Shell warned that Fortune 500 companies usually do not last more than fifty years.[11] By 2017, Credit Suisse showed that the typical lifespan of an S&P 500 company had fallen to twenty years. This is just one-third of the sixty-year average found only six decades ago.[12] What can short-lived companies like these learn from initiatives like the Olkiluoto repository project, which must maintain continuity until the facility is decommissioned around 2120? Why don't more companies have long-term safety departments like those at Posiva or its contractor Saanio & Riekkola? How can companies be made to take responsibility for environmental impacts that will outlast them?

To improve an organization's ability to zoom in and out across timescales, perhaps it could institute a days division, a decades division, a centuries division, and even a multimillennial division. The days division could focus on the organization's short-term balance sheets, inventories, human resources, branding approaches, management hierarchy, public relations, audits, and so on: the familiar hallmarks of a modern organization. The decades division, however, would explore how these strategies could later be readapted given a variety of future scenarios that could materialize over the next generation or two or three. For some, this could mean modeling future lines of evolution that society or technological innovation could follow. For others, it could mean hiring strategic foresight consultants. Such consultants already use methods with names like "alternative futures exploration." However, the forecasts would be most robust if highly trained PhDs were hired and multiple lines of reasoning were deployed. This means involving experts from several academic fields, industries, and government agencies, as they were in the Safety Case project.

Formalizing multitimescale learning by establishing days and decades divisions could be in a corporation's self-interest: it could increase its longevity. Legislation may be necessary to mandate the centuries divisions and multimillennial divisions, though. The centuries division could be required to publish long-term impact statements that aim to model, describe, or graph the organization's costs and benefits to environmental and human flourishing over, say, the next five hundred years. These could be modeled on the environmental impact statements that the US National Environmental Policy Act requires of polluting companies. Or they could resemble a Nordic nuclear waste repository project's Safety Case portfolio, except with shorter time horizons. Or they could be modeled on product lifecycle analyses. Such analyses aim to forecast a commodity's total effects on human health and the environment from its cradle (the time of its development and production) to its grave (the time of its recycling or disposal).

The multimillennial division would need to think bigger picture. It could be assisted by evolutionary biologists, archaeologists, philosophers, epidemiologists, Earth system scientists, nuclear waste experts, and/or other kinds of long-term thinkers. These experts would define exactly what, if anything, the organization can bring to the table, for our species and planet, across the coming millennia. This could mean developing something like a Safety Case. It could mean making quantitative models, qualitative analogies, or prose scenarios to envision far future impacts. Or it could mean devising a corporate "mission statement" espousing ethical goals that extend far beyond the expected lifespan of the organization itself. If the company finds itself unable to articulate its big picture contributions to humanity in a credible or convincing way, it must bear the criticism of skeptical publics and media pundits.

A GLOBAL DEEP TIME RECKONING ASSOCIATION

This chapter revealed how deep time reckoning worked best when several different kinds of experts envisioned far future worlds from several different scales, levels, and frames of analysis simultaneously. But how can we ensure that more collaborations like these come about? How can long-sighted expert teams transcend the deflation of expertise by obtaining greater prominence in the public eye? How can they gain more influence in the halls of power?

Today's deep time reckoners must organize into a more cohesive group with greater solidarity. They must forge deeper bonds based on their shared self-identifications as long-term thinkers. Climate scientists must collaborate more with nuclear waste experts. Paleontologists must talk more with archivists. Evolutionary biologists must seek greater mutual understanding with theologians who ponder God's eternity. All must commit to learning more from one another. They must work toward a more textured, multifaceted, multidimensional long-termism that defies insular information silos and disciplinary echo chambers.

To facilitate this greater cohesion, we could set up an international Deep Time Reckoning Association. This group could be modeled on highly interdisciplinary expert organizations, such as the Society for Social Studies of Science. It could be a United Nations project, an association funded by big philanthropies, or an offshoot of several professional organizations. The organizations could collaborate to establish a more broadly framed "supra-association" focused on long-term thinking. It could hold events in the spirit of the Aspen Institute: a global nonprofit think tank for idea exchange and leadership devoted to the common good. It could have a journal, conferences, a magazine, a website, and a yearly global summit. To enhance public outreach during the deflation of expertise, it could take cues from the space exploration advocacy organization, the Planetary Society. The Planetary Society is led not only by prominent astronomers and NASA scientists, but also by public icons like Bill Nye, Neil deGrasse Tyson, and Star Trek actor Robert Picardo.

One of the Deep Time Reckoning Association's goals would be to work toward incorporating more rigorous portrayals of far future worlds—seen from multiple angles, levels, and scales—into mainstream movies, television shows, podcasts, and books. For that, it could follow the lead of University of Southern California's "Hollywood, Health & Society" institute. That institute advises entertainment industry professionals on human health, security, and safety issues to improve popular media accuracy. However, the Deep Time Reckoning Association's main mission would be to nudge long-sighted experts toward self-identifying as a pragmatic multidisciplinary community that global society must rely on to avert planetary collapse. After all, only when today's deep time reckoning experts start working in tandem can a more holistic and united front against ecological and intellectual degradation take root. This means tens of thousands approaching Anthropocene problems from different time spans, perspectives, and disciplines.

PERSPECTIVE EXCHANGING PARLORS

The Deep Time Reckoning Association could set up social clubs or conference rooms where deep time reckoners from many different fields can work, socialize, and network together. Members could, for example, attend presentations inspired by the Long Now Foundation's monthly seminars, which strive to assemble a "compelling body of ideas about long-term thinking." They could discuss ideas in settings akin to large universities' alumni clubs, like the Cornell Club in New York City. Or the club could be modeled on the Cosmos Club, a private club in Washington, DC, which brings together major contributors to science, literature, public service, and art. Its membership has included three US presidents, three dozen Nobel Prize winners, sixty-one Pulitzer winners, over fifty Presidential Medal of Freedom recipients, and twelve Supreme Court justices. If a new merit-based club focused squarely on long-termist expertise were to establish its own interdisciplinary network, what sort of Anthropocene breakthroughs could emerge?

The Safety Case experts showed how cross-pollinating many different kinds of long-termist expertise can help an organization reckon deep time. Deep Time Reckoning Association members would, in a similar spirit, break down barriers between expert communities and hone long-sighted knowledge together. To facilitate this, they could do perspective-exchanging exercises modeled on Laura's forest analogy. They could routinely ask one another:

What does my work look like from your perspective? What does yours look like from mine? What does the year 22,000 CE look like from your viewpoint? What are the key phenomena we should consider? How is what you see similar or different from what I see? What about 220,000 CE? What are the blind spots in my and your expertise? How can we illuminate them for one another?

Questions like these could be posed at informal gatherings. These could resemble a more casual, inclusive version of the salons common among seventeenth- and eighteenth-century French intellectuals (and still mimicked by certain intellectual types today). There, learned people would meet to exchange ideas, tell amusing stories, and refine their knowledge and taste over snacks and drinks. The camaraderie and intellectual stimulation this offers could help deep time reckoners build community. It could give them a space to playfully think outside the box and attain a more robust, multiscale, multiperspective grasp of far futures. When events like these are not in session, the parlor could serve as a relaxed lounge or break room where deep time reckoners can achieve distance from their intense work. They could partake in counterpoint activities like playing music, painting, drawing, or reading—whatever reflects the interests of the community. They could play games likes chess, poker, Trivial Pursuit, Risk, or Settlers of Catan. This would help them take a step back from their work, as this chapter's Safety Case experts did, and sustain the aspirational motivation and intellectual energy that deep time reckoning requires.

SHIFTING ANGLES AND SCALES

Professionals from all sorts of different organizations' multimillennial divisions could pool ideas at a yearly global conference, hosted by the Deep Time Reckoning Association. Yet deep time thinking must be seen as a year-round activity, just as it was for Safety Case experts. It must be the duty of employees across entire organizations. So, when long-termist experts are struggling alone in their offices, worrying they are missing the forest for the trees, perhaps they ought to pause for a few minutes and do long-termist thought experiments. These could be inspired by how the Safety Case experts tacked back and forth across time and between others' perspectives. Introspective self-inquiries like these can cultivate experts' capacities to shift and reshift their thinking. And remember, one can engage in this activity from a wide range of backgrounds. An American farmer, for example, could do this thought experiment:

How does this year's low crop yield look from my view versus Vegetable Buyer X's view? How about US Department of Agriculture Bureaucrat Y's view? How would a locust or a crow see my situation? How do seasonal weather fluctuations affect my profit fluctuations? How would a meteorologist, ecologist, and banker see my situation differently? How is the multigenerational history of my family farm alive, or not alive, in my current way of life? How was my land's history shaped by America's homesteading settlement patterns, government programs to mitigate the Dust Bowl, Native Americans' effects on the landscape, and the extinction of North American megafauna like the mastodon? How is my agrarian cultivation of land, and my provision of food to today's populations, feeding into the long-term future of humanity?

Embarking on these sorts of multiscale, multiangle, multiperspective learning-journeys can be useful to today's deep time reckoning experts and to society as a whole. Doing personal research online or at libraries can refine one's accuracy in grasping different viewpoints, time periods, and their implications. Anyone can reinforce one's multiperspectival efforts by asking friends and coworkers lots of questions about their work's time spans—a bit like an anthropologist does. This can sharpen one's capacity to understand others' points of view. If this practice were to be widely adopted, it would not only help lay more multidimensional intellectual, institutional, and societal foundations for tackling long-term Anthropocene threats; it could also help reinvigorate enthusiasm, among skeptical members of the public, for interdisciplinary expert engagement during the deflation of expertise.

4

HOW TO FACE DEEP TIME EXPERTISE'S MORTALITY

35,012 CE: A tiny microbe floats in a large, northern lake. It does not know that the clay, silt, and mud floor below it is gaining elevation, little by little, year after year. It is unaware that, thirty millennia ago, the lake was a vast sea. Dotted with sailboats, cruise and cargo ships, it was known by humans as the Baltic. Watery straits, which connected the Baltic Sea to the North Sea, had risen above the water thousands of years ago. Denmark and Sweden fused into a single landmass. The seafloor was decompressing from the Weichselian glaciation—an enormous sheet of ice that pressed down on the land during a previous ice age. After the last man died, the landmass kept on rising. Its uplift was indifferent to human extinction. It was indifferent to how, in 2013 CE, an anthropologist and a Safety Case expert sat chatting in white chairs in Ravintola Rytmi: a cafe in Helsinki. There, the Safety Case expert relayed his projection that, by 52,000 CE, there would no longer be water separating Turku, Finland, and Stockholm, Sweden. At that point, one could walk from one city to the other on foot. The expert reckoned that, to the north—between Vaasa, Finland, and Umeå, Sweden—one would someday find a waterfall with the planet's largest deluge of flowing water. The waterfall could be found at the site of a once-submerged sea shelf. The microbe, though, does not know or care about Vaasa, Umeå, Denmark, long-lost boats, Safety Case reports, or Helsinki's past dining options. It has no concept of them. Their significances died with the humans. Nor does the microbe grasp the suffering they faced when succumbing to Anthropocene collapse. Humans' past technological feats, grand civilizations, passionate projects, intellectual triumphs, wartime sacrifices, and personal struggles are now moot. And yet, the radiological safety of the microbe's lake's waters still hinges on the work of a handful of human Safety Case experts who lived millennia ago. Thinking so far ahead, these experts never lived to see whether their deep time forecasts were accurate.

A LIVING SAFETY CASE

We have seen Safety Case experts make multiple analogies when describing their deep time reckoning projects. Some evoked imagery of living organisms and ecosystems. In chapter 3, Laura's forest analogy saw the Safety Case as a living project that required ongoing feeding. It absorbed life-giving "light" in the form of financial support from Posiva managers in and above the portfolio's "treetops." If all funding to this "forest" were to be slashed, then the Safety Case ecosystem would be deprived of its sustenance and nutrients, and it would die. For Laura, it was also the job of the experts who input data into the Safety Case's "roots" to give their "organic" collaboration the "food" it needed to grow. Another informant similarly explained her roots-level job to me as one of "feeding" the "raw" data she collects into her colleague's model so it can be "processed" and then made "ingestible" by the Safety Case as a whole. Others, as I have noted, described their long-sighted collaborative efforts as a "colony" of "ants" caught in the throes of a larger "group organism." Another scientist spoke of how the Safety Case's workings emerged, in part, "organically" through informal gossip networks and interpersonal chats among experts—through what he called *puskaradio* ("bush radio") or *viidakkorumpu* ("jungle drum").

As my time in Finland progressed, it became clear that these lively organism and ecosystem analogies were more than just figures of speech or fun expressions. The Safety Case was, after all, in a sense itself alive. It was a collaboration of living people who had to infuse their lively professional energies and intellectual vigor into the project to breathe life into it and keep it growing. But, like any living thing, the Safety Case collaboration could die. Parts of it could necrotize. Any event causing human managers, experts, or technicians to cease to pour their energies into the project could put the Safety Case portfolio at risk of withering. The most obvious examples of such life-sapping events could be a manager's unanticipated death, or the retirement of a key scientist. Or it could be an irreplaceable modeler abruptly quitting work out of frustration.

This point can be stretched further. If all of the living experts working on the Safety Case project were to simultaneously die, the portfolio would not only cease to grow, it would also lose its meaning. It would become a pile of mysterious paper artifacts decked out in barely comprehensible gibberish, strange charts, and graphs that mean little to nothing to surviving Finns. The reports would become, momentarily at least, like untranslated hieroglyphics. Unless, of course, another group of living experts were to pick up the hundreds of reports once more—putting in the interpretation work necessary to figure out the Safety Case artifact's big picture for the world.

The pragmatic, optimistic, worldly form of deep time reckoning I advocate in this book is sensitive to the possibility of death. An abrupt death of a key expert could cause, for a body of collaborating Safety Case experts, something like an organ failure. As one scientist at the Technical Research Centre of Finland (VTT) put it:

We're extracting information from old-timers. ... This is a response to a risk [of expert death] that has been realized here a number of times. ... Perhaps we should just set up some alcohol jars in the corridors and take their brains!

I broached these issues of expert mortality with my informants. Many emphasized how it can take a long time to become a top expert in the globally scarce field of repository safety assessment modeling. This steep learning curve, they told me, can make the consequences of a surprise death of an expert all the more severe. A chemist in her late fifties put it this way:

Nuclear waste is such a specialized field. At least in Finland, you cannot get enough background information at any university or in any academic course. It is something that you have to learn at the workplace. ... The process takes five or ten years, depending on the kind of work. The youngest have been here for about four to six years. You can say that the one who has been here six years has fully learned one specialized kind of work. But to get them good enough to do multiple types of jobs ... ten years.

Human mortality can be a threat to expert communities and their knowledge. It can also be a threat to societies' and institutions' capacities to think long-term. When a key deep time reckoner dies, his or her wisdom can evaporate, and the fight against the Anthropocene and the deflation of expertise can lose a key intellect. This chapter explores this challenge by telling the story of the death of a vital nuclear waste expert and its aftermath. This expert, Seppo, was described to me, with playful hyperbole, as the Safety Case's former "dictator" who "pulled all the strings" back in the 1990s and early 2000s. One insider called him Posiva's safety assessment project's Kekkonen, in a reference to Urho Kekkonen, Finland's longtime former prime minister and president from the 1950s to the 1980s. Seppo was known for his hot temper, great competence, caustic personality, salty straightforwardness, and unyielding dedication to his scientific work.

Seppo died suddenly on a Saturday evening in the summer of 2005. This dealt the Safety Case project a serious blow. The cause of Seppo's death was probably a fall on the ice. Some informants said it was a bicycle accident. Others speculated that Seppo had tripped on his shoelaces and fell. Seppo's body was found in Helsinki's Ruoholahti area near the city's *Kaapelitehdas* ("Cable Factory") building. He had recently left a graduation party at a friend's house. Paramedics tried to resuscitate him, to no avail. This tragedy brought the Safety Case, to use one informant's words, to a temporary but screeching halt. Posiva realized that the Safety Case work had come to rely so heavily on Seppo that it, as another informant put it, had become sort of a one-man show. Seppo's surviving colleagues struggled to revive their project's workflows for months. They faced thorny questions about whether it is even possible to replace highly specialized experts like Seppo.[1] From then on, some of Finland's Safety Case experts experienced "secondary haunting."[2] That is, while no informants reported literally seeing ghosts walking the halls of their workplaces, many described feeling haunted by the tragic loss of Seppo. They referenced Seppo's "specter" figuratively to underscore his ongoing impact.

In this chapter, stories of Seppo offer us a window into how the problem of expert death played out among my informants. The goal is to mine this case study for lessons about how to build more securely long-sighted institutions. Today's organizations must be more resilient to the loss of vital deep time reckoning experts. To achieve this, they must learn more about how facets of an expert's thinking and personality can disappear or persist after his or her biological death. To this end, this chapter first tracks how a dead expert's influence lived on in my informants' *predecessor parables*: the cautionary tales they told about Seppo, which conveyed lessons about how Safety Case experts ought to engage with their colleagues and their deep time knowledge. Second, the chapter delves into how Seppo's longtime "right-hand man" Gustav felt haunted by Seppo's intense personality, scientific vision, and sharp tongue. This led Gustav to, on the late Seppo's behalf, prod his colleagues to reconsider their work's direction. Studying this anthropologically revealed how lingering traces of Seppo's influence continued to infuse living experts' worlds with many different moods, debates over professional values, points of scientific debate, and invitations to reimagine how they modeled far future Finlands.

With all this in view, this chapter concludes with five more reckonings. Each explores how learning from Seppo's expertise's "afterlives" can offer lessons for tackling the challenges of replacing experts. These are challenges that commonly face rare but essential deep time reckoning specialists. In response, today's societies must embrace an ethic of *predecessor preservation*. This means carefully absorbing, tending to, and disseminating insights from prolific deep time reckoners so their contributions to humanity's long-termism do not die off upon their biological deaths. Talented long-term thinkers like Seppo merit special treatment, responsibilities, and caution from those around them. This can foster greater societal appreciation for the fragility of expert knowledge. Bringing these matters to the attention of a wider public can help us run against the grain of the deflation of expertise.

PREDECESSOR PARABLES

After Seppo's sudden death in 2005, his surviving colleagues were left confused and saddened. They scrambled to fill the leadership gaps and knowledge vacuums that his unexpected departure opened up. Some searched through folders in Seppo's personal computer for clues to his lost thinking. Others tried to interpret notes that Seppo had scribbled in the margins of earlier drafts of his reports. Posiva hired several new personnel.

Some associated Seppo's passing with the large, visible tumor on his face. Most had known that Seppo had had a vascular disease since childhood. Many had also known that he could die if the growth's blood vessels were to rupture. Seppo's health vulnerability was literally staring his colleagues in the face. Yet many still reported feeling shocked by his death. They called it "unnatural" or "untimely." One informant told me how people's tendencies to deny death can be a real liability for expert organizations. Adept at making forecasts about the Earth's radical long-term, Posiva had failed to adequately prepare itself for a completely predictable event in the radical short-term: the abrupt death of a key PhD-holding consultant who was known to have health problems. As Seneca, the Roman philosopher of Stoicism, once said two millennia ago:

We never anticipate evils before they actually arrive. ... So many funerals pass our doors, yet we never dwell on death. So many deaths are untimely, yet we make plans for our own infants: how they will don the toga, serve in the army, and succeed to their father's property.[3]

After months of workflow instability, Posiva formed the SafCa Group. This was a team of under ten specialists who collaboratively led the Safety Case project in Seppo's absence. It was overseen by Posiva's managers. This transition from single-expert to multiple-expert leadership had been underway before Seppo's death, but it accelerated greatly after it. The SafCa Group made it so that Posiva, as one informant put it, no longer "put all their eggs in one basket" with a single Safety Case knowledge chief. When I arrived in Finland in 2012, the SafCa group was in charge. Yet I was often struck by how Seppo, nearly a decade after his death, seemed to be on the tips of so many of my informants' tongues. His "specter," as some told me, still "haunted" the Safety Case expert community. As a younger Safety Case expert put it: "I've never met him, but everyone talks about him. Seppo would have said this, Seppo would have done that ... What would Seppo do here?"

Some recalled how Seppo used to storm out of meeting rooms banging doors. He only sometimes returned afterward once he cooled down. Others explained how, at meetings, Seppo was often only half following along. He rudely read through reports and listened in only when he thought something interesting was being said. He was always multitasking and looking busy. One informant described how he would fly off the handle at his secretaries. He would "directly devalue" his colleagues when he thought they were underperforming. Another speculated that Seppo, discontent with the imperfections of the world around him, yearned to live in "the perfect world of his models." One informant attributed Seppo's dogmatic tendencies to the communist leanings of his university years. He noted how, even though Seppo had abandoned his political leftism long ago, his rigid mentality toward life, work, and science had retained a fundamentalist spirit. Others associated Seppo's laconic, combative, and prideful toughness with stereotypes about Western Finland's *Pohjanmaa* (Ostrobothnia) region, where Seppo was raised. When Safety Case experts discussed these matters, they addressed how differences in individual political leanings, regional cultures, and emotional tendencies had affected their workplace relationships over the years.

Others judged Seppo's detractors to be the uptight ones. I once witnessed two Safety Case experts chuckle together, reciting one of Seppo's old jokes: "There are three types of people in this world: those who can count, and those who cannot!" Some told stories of Seppo's more jovial demeanor during sauna nights, workplace parties, or trips abroad. They recalled how Seppo enjoyed cycling and traveling across the world for vacations. The workplace, they said, was where Seppo's stubbornness, irritability, and intellectual intensity manifested most acutely. But even this intensity was thought to have its upsides. Some valorized Seppo for being so passionate and serious about his vision for the Safety Case; they found inspiration in his forefather mystique. Seppo, after all, had been a member of Finland's founding group of geologic disposal experts, nicknamed the *Paskaporukka* ("Shit Gang").

Two informants described Seppo's attitude toward Posiva as "more Popey than the Pope": more pro-Posiva than Posiva itself. Calling him a "skillful leader," another put it this way: what really angered people was that, despite being a somewhat ambiguous character, Seppo was "usually right." One cast Seppo's intellect as brilliant and his straight-to-the-point personality as charismatic. Seppo was, in these warmer recollections, summoned as an ideal role model to which today's Safety Case experts ought to conform. "Being careful was his hobby," an informant recalled, with admiration. On the other hand, memories of Seppo were also sometimes summoned to discourage peers from being nitpicking *pilkunnussijat* ("comma-fuckers")—too detail-obsessed for one's own good.

Many, though, held Seppo up as an example of how *not* to interact with one's colleagues. They spoke of Seppo's fraught working relationship with Gustav. Gustav was seen as Seppo's former lackey, henchman, or sidekick. He had a background in physics and engineering. Seppo was a systems analyst. Some cast them as what a social scientist might call a double charismatic pair.[4] Seppo was the "tyrant" with the big-picture vision; Gustav was the "right-hand man" who did the nitty-gritty calculation labor that Seppo assigned to him. An enraged Seppo fired Gustav twice. Others had similar stories. Seppo once fired Rasmus, whose modeling expertise Seppo denigrated as "like playing computer games." Both Rasmus and Gustav were promptly rehired after Seppo cooled down.

Multiple informants emphasized how Seppo had never been promoted to a management position. With many overlapping layers of authority above him, Seppo was left to micromanage his renown informally among those with whom he worked closely. Gustav told a story of how Seppo once became upset when drunk at a party. He had received news that Rasmus was promoted above him in Posiva's hierarchy. Seppo then, sadly and seriously, announced that if Gustav were ever to be promoted above him, it would be the lowest point in his life. One informant recalled how Seppo rarely talked about his private life. Another called him a lonely rider and a lone ranger. Gustav told me he had learned to keep his "personal defense lines up" when around him.

When I met Gustav, he described Seppo as an "ambiguous Angry Bird" who sometimes wore "raging bullhorns." When Safety Case experts told stories about their tense working relationship, it opened up tough conversations. They discussed how experts with controlling personalities can create difficult working environments, as they jockey to uphold their chiefly positions. Some told stories of Seppo to critique the inhumanity of experts who seem to value technical information over people. Others raised awareness about experts who seek to reinforce their mystiques as people elevated above others. Seppo's esteem, after all, relied on his standoffish charisma. Telling stories about Seppo fostered an appreciation for how, when authority is not routinized in rules—when it becomes overdependent on the potency of a single charismatic person—its leader–follower relationships may not survive the charismatic person's death.[5]

Still other Safety Case experts told tales of Seppo that warned of how experts with controlling personalities can weaken collaborations. They criticized experts who need to "hold all the reins" at work. Yearning to be the Safety Case's irreplaceable puppet master, Seppo had groomed no heirs. Instead, he hoarded his knowledge. He made himself indispensable to Posiva by strategically managing, guarding, transmitting, producing, and concealing highly coveted Safety Case information. He empowered himself by accumulating and obsessively controlling specialist knowledge that nobody else had.[6] To reinforce this, Seppo maintained a personal distance between himself and his colleagues. He ultimately became, as one informant put it, almost impossible to fire.

This set the stage for work environments that were sometimes toxic. Seppo's status enabled him to get away with being uniquely impolite, volatile, and demanding of his colleagues. Had Seppo's specialized expertise been less rare, he would have been less of a scarce human resource. Had his great competence been more replaceable, he would have had more reason to fear losing his job when he lashed out at others. But with his authority uncontested, his irascibility went unchecked. Seppo's knowledge-hoarding was enabled, in part, by his personal volatility. This volatility was itself enabled by his status; these were two sides of the same coin. As a consequence, Safety Case experts were no more likely to reminisce about the afterlives of Seppo's scientific contributions than they were about the aftershocks of his lack of collegiality. Many informants felt positive about Seppo. Others discouraged peers from regressing into outdated mentalities like his. They stressed how Seppo, while looking hundreds of millennia into the future, was a backward product of his times.

In so many fieldwork situations, Seppo's specter returned to stir up reflections on the Safety Case experts' values, relationships, and habits. My informants summoned, conjured, and channeled aspects of his past vision, character, and ideas into the present. These little everyday dramas took place over days, weeks, months, years, and decades, yet they affected how Posiva reckoned multimillennial futures. Stories about Seppo were derived from the past, yet they were told to convey value judgments about the present. Since Seppo's death, his surviving colleagues have shared cautionary tales about how uneven distributions of unique expert knowledge can become liabilities for organizations that rely too heavily on a single specialist. They referenced Seppo when sharing opinions about favorable versus unfavorable, and polite versus impolite, styles of expert behavior among the living. This was just one way that traces of Seppo's past life influenced the Safety Case experts' professional community.

Studying these predecessor parables anthropologically revealed how aspects of a lost expert can, in philosopher Søren Kierkegaard's terms, be "recollected forward" from the past into the present.[7] These aspects included Seppo's past emotions, technical knowledge, dreams, personality quirks, ideas, visions of the future, and more. When summoning Seppo's specters, informants often characterized him as strong-willed, competent, and reputable—but also intense and best kept at arm's length. This reinforced Seppo's role, in death as in life, as what anthropologists call an exemplar. He was a figure placed on a pedestal. He shaped his community's imaginations, values, and ways.[8]

Telling stories of Seppo, my informants taught lessons about expert collegiality, authority, vision, temperament, passion, charisma, kindness, politeness, social distance, mystique, and leadership. They did so with solemnity, awe, respect, nostalgia, aversion, or warm friendliness, for different reasons at different moments. Seppo's successors curated memories of their lost forerunner. They empowered Seppo as an influential predecessor. Recollections of him intervened in the Safety Case experts' worlds, opening his surviving colleagues up to negotiating and renegotiating how they related to one another and to their scientific work. In these ways, remnants of Seppo's past impact lived on to subtly affect the organization of Posiva's Safety Case collaborations, influencing how their visions of far future Finlands took shape.

SPECTERS OF SEPPO

When I met with Gustav in February 2013, he showed up mildly hungover. He had two candy bars, a pack of cigarettes, and a few cups worth of coffee in hand. He complained of working long hours; he was tired of recalculating his datasets every time new information was found or flashy new computer programs were released. He griped about having to leave his *kesämökki* in mid-July to recalculate his radionuclide transport models, and about his workplace's new scientifically inexperienced "business-oriented" managers. He criticized Posiva's pressures to write lengthy Safety Case reports in "Oxford English." He mocked a colleague who had recently denigrated his work as "only mechanical calculation." Sitting in front of his computer, Gustav walked me through why crunching chemical inventory datasets on today's "cutting edge" software was clumsy compared to the simpler, more straightforward UNIX software of the 1990s. He was fed up, as Seppo used to be, with scientific research's twenty-first-century corporatization. Gustav was also no longer so big a "fan" of the USA, either. He had stopped voting for Finland's conservative party *Kokoomus*. He had begun voting for Social Democratic politicians and, as Seppo once did, even a Left Alliance candidate.

These days, Gustav insisted, the Safety Case community is "all about market economics and competition," with everyone trying to "advertise" their expertise to everyone else. He was displeased at how many experts today feel that their own work is of the utmost importance. If Seppo were still alive, Gustav explained, he too would dislike these trends. In the 1980s, Finland's nuclear waste expert community felt, to Gustav, more "like brothers" or a "big family." In his eyes, it used to be a band of "crusaders" working toward a "good and honest safety assessment" and nothing else. Gustav felt sick when he heard rumors of fellow VTT colleagues hiding research findings from one another, seeking only to advance their own careers. He was tired of today's Safety Case experts self-servingly arguing for extensions to Posiva's funding contracts. Their goal, he said, was to "keep the problem alive" by claiming again and again that still more research had to be done. Gustav lamented how the "spirit of our times" is "formalist" and "cosmetological"—lacking in substance.

Gustav once described Seppo as ugly. Another time he compared him to Stalin. But he missed Seppo's willpower in "standing up" to the private consultants who, as he saw it, sought never-ending funding from Posiva. Before he died, Seppo was irked by Posiva's decision to start modeling highly unlikely scenarios—such as the total failure of a copper canister, or the complete disintegration of a bentonite clay buffer. Seppo was displeased with how Posiva approached its transition from the 1990s' static safety assessment models to the early 2000s' time-dependent models. He spent his final years insisting that the models must be kept simple enough for a single talented scientific brain (like his) to understand them in their entirety. Like Seppo, Gustav did not want the Safety Case to become more bloated, costly, or mucked up by excessive documentation rules. One colleague noted how Seppo believed that the nuclear waste problem had already been solved. The only task left was to write reports explaining to "all the mentally retarded people" why Finland's and Sweden's repository designs made it a "piece of cake." As Seppo's computer screensaver used to read: "No More Research Is Needed."

Gustav told me how Seppo's past vision still inspires him to "be more aggressive with bullshit," helping him muster the courage to speak out against those who he saw as degenerating the Safety Case. After one contentious meeting, a colleague looked to Gustav and asked, "Are you missing Seppo right now?" Gustav responded by criticizing the SafCa Group's new corporate-managerial spirit. He cited Seppo's "No More Research Is Needed" mantra. In these moments, Seppo's specter drew Gustav to lobby for "another way—an old way, but also, potentially, an alternative way—of doing things."[9] Seppo's vision became, for Gustav, an inspiration for prodding his colleagues, on his late boss's behalf, to rethink the Safety Case's direction. Through Gustav, Seppo's influence helped steer Posiva's deep time reckoning efforts, even after his passing.

Gustav still felt Seppo's oversight over him. Once he half-joked to me that he sometimes imagined his late boss, sitting on a cloud in the sky, disgusted with the SafCa Group's efforts. In Gustav's daydream, Seppo begs God, "Please, give me a pistol, I will shoot myself!" God replies, "Seppo, you are already dead; that will not work." Seppo then pauses and exclaims, "Then send me to Hell so I cannot watch this anymore." Seppo had, to use the philosopher Jacques Derrida's terms, a powerful "visor effect" over Gustav. He had feelings of being watched by Seppo's judgmental gaze without being able to watch Seppo back.[10] The late systems analyst's watchful eye pressured him to perform. It controlled him even in death. Seppo's influence left Gustav with a feeling of something-to-be-done[11]—a nagging "this aspect of this dataset on Olkiluoto's hydrogeology needs to be more like that" or a judgmental "today's models are unnecessarily complex." These somethings-to-be-done called on Gustav to pursue his work in alternative ways. Seppo's past vision for the Safety

Case thereby shaped Gustav's relationships with other experts, as well as his scientific work.

Echoes of Seppo's sharp tongue and angry outbursts still rang in Gustav's ears. He once loudly mimicked Seppo's hollering "ARAGHGHH!!!!" when relating how Seppo in 1992 furiously told him that their scenarios about far future postglacial periods were "rubbish." Gustav told Seppo he had thought the same thing, but that Seppo "was not the type of person you give advice to." Gustav then did more supercomputer tests and printed out his findings. Seppo responded "ARAGHGHH!!!!" Another time, Gustav reenacted for me Seppo's rage in 1995 after he discovered that Gustav had requested data from Finland's state Game and Fishing Services unit without his permission. Gustav had been skeptical about the accuracy of certain modeling assumptions they had made about Finland's far future fish and animal populations. Seppo angrily shouted at him, "Gustav, your job is to solve problems, not create them!" When I met him years later, Gustav still felt his workplace emotions and moods periodically agitated by the spell of Seppo's past oversight. Colleagues joked that Gustav "deserves a medal" for working with Seppo for so long.

Yet Seppo haunted Gustav as both the cause and the solution to his problems. He channeled his late boss to both initiate and resolve workplace conflicts. Seppo was seen as intensely rational; he used to propose that, when scientific arguments begin to get out of hand, everyone should "put it on paper" before debating further. Writing down one's logic and calculations can give each party time to clarify their positions and discover potential errors in their reasoning. After he died, Gustav started advocating this same practice among those working for him. He drew on Seppo's scientific approach to try to elevate the Safety Case work, and the discussions surrounding it, to what he saw as a higher ground. Sometimes Gustav's sentiments toward Seppo resembled the great ambivalence, documented by many anthropologists, that many of us feel toward the influence our ancestors, forebears, or dead elders have over us.[12] Other times I worried that Gustav was weighed down by what philosopher Friedrich Nietzsche called *ressentiment*.[13] He seemed mired in the will of an idol-adversary from a frozen past. What was clear, though, was that many of the 2010s SafCa Group reforms that Gustav loathed were, ironically, sparked in part by lessons that Posiva learned from Seppo's passing.

As the previous section explained, the predecessor parables that Safety Case experts told about Seppo's death nudged Posiva to reconsider the risks associated with relying too much on a small handful of highly specialized mortal experts. This was seen as especially important when dealing with experts, like Seppo, who were reluctant to document the assumptions and methods underlying their calculations. Posiva responded to this by upping its documentation requirements and "knowledge management" oversight programs. This meant accelerating and expanding certain reforms that Seppo himself had, ironically, long despised. After Seppo passed away, Posiva started devoting entire sections of its reports to "transparently" describing its own uncertainties, "traceably" outlining its methodologies, and "conservatively" formulating many possible future scenarios. This included extremely pessimistic and unlikely scenarios. Posiva released new reports like *Models and Data*, which summarized how various models and datasets were made, how they linked together, and how their knowledge gaps could be filled. One informant called *Models and Data* the Safety Case's "phonebook." Its goal was to present both knowledge and uncertainties in accessible formats for Posiva's many stakeholders. These stakeholders included STUK regulators, the Eurajoki community, international reviewers, the Finnish public, politicians, and civil servant experts in government ministries.

Gustav disliked this new paradigm. But it was in part inspired by memories of Seppo's knowledge-hoarding. When Seppo died, Finland's nuclear waste repository safety assessors' old guard lost a fierce advocate. Gustav gritted his teeth as projects on Olkiluoto's far future biosphere "exploded" with what he saw as expensive, unnecessary, excessive reportage. The biosphere models changed substantively too. Seppo's 1999 safety assessment report presented models of generic ecosystems that could exist in many other locations globally. He used to criticize costly proposals to develop ecosystems models that are far more specific to Olkiluoto. Yet, by the time I began fieldwork, biosphere assessors were measuring local fish, monitoring the area's forests, sonar-scanning sea bottoms, simulating Finland's changing shoreline, and plotting out a food web of over fifty regional species. With Seppo no longer alive, Gustav lacked a strong ally to help him oppose these reforms.

At the same time, traces of Seppo's past were also working against Gustav's biosphere modeling agenda. Seppo's volatility was part of what initiated the biosphere model's transformation in the first place. Juha, a Safety Case expert in his thirties, described his early-2000s work to tirelessly master the Safety Case's biosphere model as a reaction to Seppo's inability to accept criticism. Seppo used to periodically highlight Juha's inferior understanding of the safety assessment project as a whole. Juha's tactic was to counter him by highlighting Seppo's inferior understanding of its biosphere section. As time went on, Juha went "deeper and deeper into the mire." He perfected the biosphere project as his own scientific "playground" where he had an edge over Seppo, not vice versa.

Today, Posiva's biosphere model is expansive, complex, time-dependent, and site-specific. This complexity had its origins in a passion project that Juha hatched, in part, as a personal defense against Seppo's irascibility. When Seppo died, room opened up for Juha to further assert his expertise. Gustav, still haunted by Seppo, would sometimes send barbs his way. But the barbs only motivated Juha to go deeper into the mire. With Seppo's past derision haunting Juha, he worked intensely to expand the biosphere project. Some grew concerned about occupational burnout. Juha's great ambition began to irk Gustav: a man who had adopted many aspects of the mentality of the man it was originally intended to irk. Trapped in Seppo's vision and oversight—as both the reason and remedy for his workplace discontents—Gustav had not escaped Seppo's hold over him.

AFTERLIVES OF EXPERTISE

Seppo's story, at first blush, may seem so unique to Finland's nuclear waste expert worlds that its relevance is limited to that community. This is not the case. Nuclear waste, weapons, and energy experts across North America, Western Europe, East Asia, and beyond have grappled with the problem of expert mortality. During my fieldwork, they were all in the midst of mass baby boom generation retirements and intergenerational transitions. This sparked widespread concerns about how unplanned-for retirement, outsourcing, downsizing, job transfer, death, or quitting can increase project instability. This instability can be amplified when an expert with "valuable and unique knowledge"—a "go-to" person like Seppo who "peers and management recognize" as someone "we can least afford to lose"[14]—is lost.

As an example, after American nuclear weapons designer Seymour Sack's 1990 retirement and 2011 death, colleagues worried his departure could pose serious knowledge loss problems at Lawrence Livermore National Laboratory.[15] Working in a context of national security secrecy, Sack mentored primarily through personal relationships. He did not leave behind detailed reports on his work.[16] Like Seppo, Sack was hardly known outside his small circles of experts, but was renowned within them for his great competence, bluntness, and gruffness. He was a mortal beacon of scarce, rarefied, coveted nuclear expertise. Precautions had to be taken to ensure his knowhow outlasted his body's death. Sack built nuclear weapons that could cause mass species extinction and irreversibly change the Earth's environmental and human futures. Yet his fragile knowledge remained vulnerable to the fleetingness of a single human life.

The challenges that mortality poses to knowledge are alive outside nuclear communities as well. Corporate "key person" or "key employee" insurance plans are testament to this.[17] A Finnish physicist once told me about a VTT computer scientist who was the only one who knew how to use a certain computer code. Before dying suddenly, he had not documented his work so "an outsider could continue with it." This left his surviving colleagues with no choice but to abandon it and start developing a new code. They accepted "at least twenty years of experience wasted." Similarly, when trying to get financial systems running again after 9/11, traders from a Manhattan securities firm strained to recall the names of deceased colleagues' family members and vacation destinations. Their goal was to guess computer passwords that went to the grave with them.[18] A program manager from a large American conservation NGO once told me a similar story about a state-level director who died suddenly in an avalanche. After the tragedy, it was difficult to continue a key real estate project without his knowledge of the deal's moving parts and many relationships in the region. Unanticipated deaths like these can destabilize any corporate, government, or academic organization reliant on small teams of narrowly specialized experts who harbor rare-but-coveted knowledge that takes years to acquire.

Fortunately, however, all aspects of an expert's, scientist's, or intellectual's contributions do not vanish upon his or her biological death. Seppo's story was evidence of this. This is a phenomenon that anthropologists have long appreciated. Gregory Bateson once noted how, while Socrates the "bioenergetic individual" is dead, much of him still lives on to shape a "contemporary ecology of ideas."[19] Bruno Latour once argued that he dwells in the "continued" history of Louis Pasteur's "network" every time he eats pasteurized yogurt.[20] More recently, Dominic Boyer has explored how, although German philosophers such as Hegel may be dead, aspects of their thought are still alive among Germany's media intellectuals.[21]

Seppo, of course, was not a famous theorist from decades, centuries, or millennia past. Nor was he globally renowned for introducing any grand idea, theory, or invention. But for the small teams working with him, he was indispensable. This made his influence outlast his body's death. Many told predecessor parables about him. Some felt haunted by him. Remnants of his thinking patterns, his emotional outbursts, his outsized personality, his prolific skill, his vast knowledge, and his defiance of expert norms of politeness had powerful afterlives. These afterlives, my fieldwork experiences taught me, can carry with them useful lessons. They can inform other places where highly specialized knowledge becomes so valuable that it warrants meticulous preservation. This brings us back to this book's main goal: to learn how to better cultivate, conserve, and circulate long-termist thought in order to mitigate the Anthropocene and the deflation of expertise.

PREDECESSOR PRESERVATION

Deep time knowledge is vulnerable to the cessation of human life. It is alive, fragile, and scarce; it often takes decades to acquire. Societies seeking to preserve their deep time reckoning human capital can read stories like Seppo's as cautionary tales. These stories can teach us about the challenge of supporting rare but essential teams of long-sighted experts who can lengthen our thinking's timescales during the Anthropocene.

This book's introduction criticized today's societies' meager reserves of highly trained long-term thinkers. It warned of a growing crisis of deflated enthusiasm for and lapsed faith in expert knowledge. It argued that these tides must be reversed if we are to avert planetary destruction. This must be done in a careful, practical, self-critical way. Learning from Seppo's story can inspire a path forward: a vision of a world in which just one talented, passionate, devoted expert becomes remarkably valuable to a community that makes deep time learning its priority. This inclination to value, rely on, and preserve long-termist knowledge need not be local to an elite few in Europe's far north. Appreciating the indispensability of skilled deep time reckoners like Seppo must be our grandest societal aspiration during the deflation of expertise. It must become a collective sensibility that we, as denizens of the Anthropocene, strive to attain. At the same time, however, Seppo's story can also inspire us to more strategically mitigate any unsavory personality tendencies at work. This is crucial when deep time reckoners find themselves dealing with exceptionally talented colleagues who happen to be as irascible and combative as Seppo was. This chapter has also shown how toxic workplace relationships can obstruct long-termism.

Communities reliant on slow-to-acquire, sophisticated, scarce expert knowledge can be sources of learning as we work toward building more long-sighted societies. Many lessons can be gleaned from Seppo's death. Seppo's context was, after all, one in which the multimillennial time spans of nuclear waste risk, the intergenerational time spans of expert succession planning, and the fleeting time spans of a single expert life shared entwined fates. This temporal mashup has great relevance to the Anthropocene's tangles of human and geological timescales. So, let's scour the promises and perils of Seppo's story for reckonings that can help expert communities better hedge against the risks of knowledge loss that expert mortality can pose.

The reckonings that follow ask: How can we brainstorm new programs and norms of professional engagement among communities of deep time reckoners—be they climatologists, paleontologists, philosophers, nuclear waste experts, biodiversity specialists, evolutionary biologists, historians, archaeologists, astrophysicists, or others—plus those working closely with them? Can tomorrow's institutions more effectively preserve mortal, fragile deep time knowledge? How can organizations learn to better identify prolific deep time reckoners in their midst? How can the public realize these skills' importance to the future of humanity? Must deep time reckoners receive special treatment that both grants them extra privileges and requires of them extra responsibilities? How can societies provide more positive reinforcement to support long-sighted experts, given how alien this supportiveness would be to today's deflation of expertise?

RECKONINGS

MULTIPLE EGG BASKETS

Seppo's death has shown why a deep time reckoning project with unique, irreplaceable personnel should never put all its eggs in one basket of a single leader or hotshot specialist. A multiexpert team like the SafCa Group can, we have seen, offer a better hedge against death's unpredictability. It can distribute irreplaceable long-termist knowledge more widely by not clustering it into a centralized "choke point" person who is susceptible to untimely death.[22] Given how valuable today's deep time reckoners' knowledge is to society at large, we should press the organizations that employ them to implement protocols requiring more decentralized networks of leaders.

Key deep time reckoners like Seppo ought to be strategically split up—put to work in separate offices or laboratories. Doing so could prevent irreversible damage to societies' long-term thinking capacities if a catastrophe were to strike a location that would otherwise result in simultaneous group death among key deep time reckoners. Such crucial long-sighted teams should also never all fly on the same plane or ride in the same van or bus together. If they do, a single transportation accident could erase their deep time knowhow all at once. If key academic, corporate, or bureaucratic organizations are not interested in implementing these policies specific to deep time reckoners on their own, then voters could push for laws requiring them. After all, many top corporations limit the number of executives they allow to be on any given flight. Are CEOs, COOs, and CFOs really more valuable to society than highly trained deep time reckoners? These are, after all, the intellects we need most to save us from Anthropocene collapse.

UNHOARDING KNOWLEDGE

A key expert's volatile, controlling, or autocratic personality may be efficient in the short term. But it can threaten the long-term preservability of his or her knowledge. When vital long-sighted knowledge is at stake, colleagues must openly discuss personality volatility risks. This can be done in meetings behind the expert's back at first, but eventually it must be to his or her face. Institutions must also develop policies for ensuring this expert cannot imprison his or her most valuable deep time knowledge in the solitary confinement of a single mortal brain.

There are plenty of precedents for policies like these in nuclear industry "knowledge management" programs.[23] Such programs can be mined for techniques that can be adopted by organizations that employ long-term thinkers. For example, any talented deep time reckoner should document his or her knowledge in reports. He or she should groom protégés as successors. He or she should communicate openly with deputy experts in line to take over as replacements if he or she unexpectedly dies, retires, or leaves work. On-the-job mentoring relationships and face-to-face apprenticeships are needed to transmit long-termist intuitions to new generations. All deep time wisdom ought to be stored in databases too, with backups secured in multiple locales across the globe. Perhaps these storage systems could be informed by Finland's nuclear regulator STUK's long-term archives.

PREDECESSOR PRESERVATION

Seppo had some personality flaws; but he also showed how a strong-willed, talented, hardworking expert can be so influential, productive, and innovative that he or she brings more to the table than a whole handful of less passionate, less meticulous, follow-the-leader experts can. These prolific experts must be identified and made aware of the responsibilities that their privileged status entails.

In Sweden's nuclear waste program, key soon-to-retire experts have been interviewed extensively to record and archive testaments. These testaments have detailed their knowledge and careers for the future experts who will succeed them. This is a good start; but, with a matter as grave as today's need to better preserve deep time wisdom for humanity, it falls short. For one, it focuses too narrowly on *key individuals* and not the *wider network* of colleagues, ideas, technologies, administrative supports, funding streams, and scientific trainings that helped create, cultivate, and contribute to the exceptional individual's achievements. To even scratch the surface of archiving a prolific deep time reckoner's lifetime impact, dozens and dozens of associated colleagues and coworkers must be interviewed about the expert's thinking. This is something I learned gradually as I conducted interview after interview about Seppo. To this end, organizations employing deep time reckoners can take cues from nineteenth- and twentieth-century "salvage anthropology" projects. These were efforts to collect artifacts, archive knowledge, and interview insiders of cultural groups with dwindling numbers. The aspiration was to preserve some of their precious ways and ideas before their last surviving members died or assimilated into other societies.

GLOBAL DEEP TIME RECKONING
INFORMATION REPOSITORY

A key deep time reckoner's knowledge must not be hoarded by any individual or organization. It must be conserved for posterity and disseminated widely across populations. Perhaps, then, the findings, forecasts, scenarios, ideas, models, interview recordings, meetings transcripts, and notebooks from deep time reckoners from many different sectors and fields should be archived. Their long-termist knowledge could be databased and backed up in an international information repository.

This global database could be made fully transparent online, just like Posiva's information Databank or NAWG's natural analogues Library archive. It would help communities of long-termist experts who are not usually in contact with one another to cross-pollinate their farsighted ideas. Each entry would also include an educational, plain-language description to help concerned citizens use it as a resource for their own long-termist learning. This would enable deep time knowledge to serve all sorts of alternative purposes not specific to the organizations that produced it (similar to how I am, in this book, repurposing a nuclear waste organization's knowledge for anthropological purposes). The database could perhaps be overseen by UNESCO or a trusted NGO. Inspiration for it could be drawn from the nuclear energy industry's worldwide failure analysis databases: searchable systems that collect and catalog malfunctions, accidents, and other learning-experience insights. These databases compile reports from many nuclear facilities across the world to assist future generations if similar problems arise later on.

AFTERLIVES OF EXPERTISE

Alongside new programs and policies, subtle shifts in deep time reckoners' workplace cultures are necessary. While a predecessor expert (like Seppo) may be dead, that does not mean he or she is no longer shaping the knowledge, relationships, and values of successor experts (like Gustav, Rasmus, or Juha). As Gustav's reassertion of Seppo's "No More Research Is Needed" slogan demonstrated, predecessor experts' past visions of the future can be usefully summoned into the present. This can shake up colleagues' preconceptions about how to proceed with their scientific efforts, and it can introduce alternative perspectives into workplace debates.

Long-sighted institutions need to appreciate, as anthropologists long have, that a person's biological death is not the same as a person's social death.[24] In the wake of the abrupt loss of an expert, surviving colleagues should be encouraged to pause and reflect. They could ask themselves: what aspects of the deceased expert's knowledge and personality still live on through me and my work? Alternatively, they could set up workplace meetings in which experts self-scrutinize the parts of themselves that might be reflections of, contributions from, defense mechanisms against, or inspired by those of a lost deep time reckoner. Doing this in a group may be too difficult given the emotionally taxing nature of the tragic situation. In that case, experts could be encouraged to meditate on how a predecessor's wisdom or weaknesses may still linger on in them. Personal intellectual exercises like these can provide surviving experts with a richer sense of how traces of the late expert can be most usefully channeled, conjured, or summoned during future troubleshooting moments.

SUCCESSOR STEWARDSHIP

Efforts to preserve vital patterns of long-term thinking must be supplemented with something more robust than formal knowledge management programs or databases. After all, what Seppo's surviving colleagues retained from their past mentor included much more than the mere storage of technical information or the backing up of scientific findings. It spanned Seppo's whole gamut of emotions, passions, relationships, personality quirks, and flaws. Traces of Seppo's total person lingered on to shape my informants' everyday office lives. This included damages that his sharp tongue and volatile temperament did to working relationships. Gustav, for example, was both inspired and haunted by Seppo's ambiguous character. This affected how he performed, interacted, and thought, for many years after Seppo's death.

Keeping these possibilities in mind, successor experts should be encouraged to reflect on which aspects of a past deep time reckoner were ultimately helpful versus damaging to their organization's long-termism. Then they could work to overcome the predecessor's faults by making mature determinations about which aspects of the lost expert to inherit or not inherit. This means adopting an ethic of succession stewardship: of responsibly selecting which aspects of a predecessor should be mimicked and replicated versus which should be discouraged and discarded. Each member of each new generation must see him- or herself as both (a) a mouthpiece for transmitting predecessors' long-term thinking patterns and (b) a reformer ending the transmission of impediments to this long-termism. Adopting this principled ethic or code of action is essential to ensuring that experts' deep time reckoning skills are sustained and refined across the generations.

CONCLUSION
Escaping Shallow Time Discipline

This book's introduction described a host of powerful cultural, economic, political, and technological forces that today, unfortunately, trap billions of minds in the narrowness of the now. Imposing *shallow time discipline* across populations, these forces of short-termism may seem so deeply rooted that, when reformers call for change, our guts tell us progress is impossible. These forces may feel so familiar that, when critics scorn the risks they pose, it begins to sound like a stale cliché. This, however, does not mean that our mission should be abandoned; it simply underscores the problem's prevalence. The past four chapters have shown that today's societies, fortunately, already have many little-known, personally enriching, practical ways of resisting shallow time discipline at their fingertips. Our challenge now is to discover creative ways of learning from these expertise-driven long-termisms during the Anthropocene. It is also to help them gain greater visibility across society, media outlets, and political debates during the deflation of expertise.

In chapter 1, we learned how Safety Case experts drew on archaeological and geological evidence to make analogies that stretch widely across distant times and places. In chapter 2, we learned how they drew on simple, repeating, organizing patterns like input/output to help make quantitative models of far future Finlands. These chapters closed with reckonings that took the approach of "How can we begin changing the world by changing ourselves?" They brainstormed starting points for long-sighted thought experiments and open-minded intellectual attitudes that any of us can adopt. In chapter 3, we learned how Safety Case experts zoomed in and out on visions of deep time from multiple angles, perspectives, and scales. In chapter 4, we learned how they grappled with the untimely death of a crucial deep time reckoner, plus the personnel- and knowledge-loss consequences it triggered. These chapters closed with reckonings that took the approach of "How can we change society by reforming how today's thought-leaders operate?" Suggesting that we bring long-termist expertise closer to the center of societal decision-making, they asked us to consider possible transformations in how institutions, experts, and societies organize their deep time reckoning projects.

Together, these strategies can inch us closer to achieving our long-termist goals. Unfortunately, though, one anthropologist's long-termism would never be enough. Nor would that of a few dozen Safety Case experts from Northern Europe. Entire populations must commit to pursuing adventurous, ongoing, long-termist learning. Individuals must voluntarily take up this mission in whatever ways best suit their unique personalities, circumstances, and talents. Bjornerud has argued that "building a more robust, enlightened, time-literate society" that makes "decisions on inter-generational timescales" requires a shift in our very perception of time.[1] This is likely so. To survive the Anthropocene, billions of us must work toward widening our intellects' time horizons. Changes are needed at the societal level and on the global scale. We must learn to reform our basic intuitions and rise above the short-termist incentives before us. We must resist the deflation of expertise and seek out highly trained deep time reckoners in our midst. Those in positions to push for systemic change in today's centers of power must also take up this cause.

Of course, this is a tall order. But if this thought-revolution sounds absurd, I reply: is it really any less absurd than staying on our current course, careening toward an Anthropocene cliff?

With these challenges in view, we can close by drawing on the case studies and reckonings of chapters 1 through 4—pulling them together and setting them into motion, with an eye to the future. In this spirit, I present two final thought experiments. Both, in chapter 3's terms, take a step back from, zoom out from, or think bigger-picture about where this book's reckonings could lead us. The first asks if what we need is a radically new education program. This program would introduce entire populations to humanity's inventory of long-term thinking techniques from an early age. Works like *Deep Time Reckoning* would be but one assignment in a broader, more long-sighted curriculum. How would societies' decision-making customs change if, before reaching adulthood, everyone had already considered, debated, and digested dozens and dozens of tools for long-term thinking, including the Safety Case experts' analogue, modeling, and scenario forecasting methods? The second thought experiment zooms out even further to ask: How could a hypothetical society, already steeped in long-termist learning, be organized differently? How could it better override shallow time discipline?

Taking these two hypothetical worlds as inspirations, we can then reflect on how to overcome rampant shortsightedness before any hopes for human flourishing succumb to the Anthropocene and the deflation of expertise.

ANTHROPOCENE CITIZENSHIP 101

In 2013, SpaceX and Tesla Motors CEO Elon Musk asked: "Where will we be in one hundred thousand years if we continue anywhere near our current pace?" Musk answered himself: "Either extinct or on a lot of planets." This may sound like a bit of a gimmicky question. It may sound like a quirky way to jog the imagination—or the oddball musings of a science fiction enthusiast. Or maybe it sounds like the out-of-touch fantasy of a pompous tech billionaire, seeing himself as an elite vanguard of innovation, shepherding humanity into the future. But what if we had been taught, from an early age, that Musk's query is one of the most important questions we can ever ask ourselves? What if we truly believed it was *our* job, and not just some eccentric billionaire's job, to answer it? What if we had been raised to see long-term thinking as our civic duty—told that, like avoiding littering, it is essential to responsible citizenship? What if our journey of long-termist learning had begun at a much earlier stage of our development into adults?

Let's imagine we are in junior high school. We are not only taking a history class, but also a deep history class and a futures class. These are part of a multiyear Anthropocene civics course that begins in elementary school and continues through high school. We go on field trips to local geological formations to help us make deep time analogies like those found in chapter 1. We do homework assignments stringing together threads of futurological thought, using input/output and if/then thinking patterns like those examined in chapter 2. We spend weeks learning about the Big Bang 13.8 billion years ago, the formation of the Earth 4.5 billion years ago, the rise of life, the age of the dinosaurs, the evolution of *Homo sapiens*, humanity's migrations across the continents, the causes of the Anthropocene, the history of technology, and more. We learn about Thomas Edison's 1911 prediction that, liberated from poverty in 2011, we would see houses furnished top-to-bottom with ultra-lightweight steel, including steel baby cribs and dinner tables. We learn about economist John Maynard Keynes's 1930 prediction that technological progress would deliver advanced economies fifteen-hour work weeks by the year 2000.[2] We learn about other early- and mid-twentieth century unfulfilled predictions that, today, we would see flying cars, immortality pills, force fields, colonies on Mars, and teleportation devices.[3] Then we learn about the many economic, cultural, scientific, and historical reasons that none of these visions come to fruition.

We learn about today's Svalbard Global Seed Vault project to back up plant seeds in an Arctic gene-bank repository to ensure against genetic diversity loss during future global crises. We learn about climate modelers' visions of future ecosystems, about the US WIPP nuclear waste repository's proposed warning monuments, and about indigenous peoples' "distant time stories" of ancestral pasts.[4] We learn about how human language has changed across millennia, plus the world origin stories of Christianity, Islam, Hinduism, and other traditions. We learn about cosmological assumptions made by past societies—from the ancient Egyptians to the Aztecs to Shang Dynasty China. We reflect on what it must have been like to see the world through their eyes.

Discussing how Earth's population could grow to nine or ten billion by 2050, we speculate about what it might be in 102,019 CE. We reflect on notions like the ancient Norse *wyrd*, which is about appreciating the past's presence in the present.[5] We ask how these notions, when coupled with geological knowledge, can help us see how the distant past is still alive in today's rocks, glaciers, ecosystems, and terrains. We examine dozens of other long-termist case studies, and we write detailed essays about them, called "reckonings."

Now let's imagine we are in high school. We watch short films on multitimescale awareness released by experts from the Global Deep Time Reckoning Association proposed in chapter 3. We write essays critically analyzing influential renderings of the future. These include Ray Kurzweil's vision of the Singularity,[6] Marxists' past visions of coming communist utopias,[7] and Alan Weisman's *The World without Us*.[8] We take notes on astrobiologist Lewis Dartnell's how-to manual for restarting human progress after global collapse, titled *The Knowledge: How to Rebuild Our World from Scratch*.[9] We write book reviews of post-apocalyptic science fiction novels like Walter Miller Jr.'s *A Canticle for Leibowitz*.[10] We are assigned textbooks showcasing long-termist tools stored for posterity in the Global Deep Time Reckoning Information Repository proposed in chapter 4.

We learn about the tools that actuaries and underwriters in the reinsurance industry use when trying to, decades in advance, shield insurance companies from insolvency due to future calamities like hurricanes, earthquakes, or wars. We learn about techniques the Catholic church has developed to transmit messages, archive information, oversee property, and maintain institutional continuity across centuries and millennia. We learn about the deep time reckoning techniques developed in the nuclear waste management programs of Canada, Switzerland, France, Germany, Japan, the United Kingdom, and the United States.

We learn about problems of knowledge loss posed by digital obsolescence: when a digital file is no longer openable because of outdated software or bygone hardware needed to run it. We learn how philanthropic organizations, run by the ultra-rich, plan humanitarian projects in multidecade timescales. We also learn about how some see their special privilege to do so as plutocratic. We learn how different languages have different ways of articulating time. Finnish, for instance, has no future tense. We learn about different cultural attitudes toward futures: a Finnish informant's favorite phrase *pessimisti ei pety* ("A pessimist is not disappointed") comes to mind. We discuss how future cyborg technologies and genetic modification techniques could alter humans' physiologies and mental abilities. We ask if this could create even more inequalities if the enhancements are not shared across society. We debate when and if robots and artificial intelligence will supersede organic human intelligence. We consider the Future of Life Institute's work to highlight the risks that artificial intelligence may pose. We examine dozens of other long-termist case studies too, again writing about them in term paper essays called "reckonings."

One principle guiding our education would be ecologist Aldo Leopold's call to think about the living world more holistically "like a mountain."[11] A second would be Finland's Safety Case experts' principle of interdisciplinary inquiry, known as "multiple lines of reasoning." A third would be legal scholar Frank Pasquale's principle of "attunement": an embrace of receptivity, sensitivity, and appreciation of one's wider world as opposed to yearning for mastery, escapism, and mechanistic thinking. Pasquale's attunement principle's spirit of self-searching was inspired by Pope Francis's "sensitivity to time and speed."[12] A fourth principle would be this book's contention that, during the Anthropocene, everyone has something to teach and something to learn. A fifth would be the view that "imagining diverse futures" can "reveal unseen pathways and can inspire human ingenuity" while also revealing the "limitations of human agency in a complex world."[13]

With these pillars instilled in us, we would read case studies in building more long-sighted societies. We would learn how certain hunting communities have rich knowledge of long-term wildlife population management and sustainable hunting practices. We would learn about inner-city communities that have insights into gentrification and urban planning futures. We would learn how certain evangelicals have developed subtle ways of navigating the long-term past and future narratives presented in both Christian creationist theories and secular biological evolution theories. We would examine dozens of other communities as well—widening not only our time horizons, but also our vision of humanity.

We would explore barriers to long-term thinking. We would examine how humans and their biological ancestors have, over millions and millions of years, evolved to react mostly to short-term crises (think lion attacks, droughts, warfare raids, feuds in one's community, food shortages, providing for one's kin, competition for social status, sex, and reproduction). We would reflect on how environmental disasters looming two centuries from now—like those considered in the Safety Case—rarely send people's fight-or-flight reflexes into panic mode. We would ask why appeals to "slow violence" often fail to win hearts and minds or incite political action in today's spectacle-driven media landscape.[14] Slow violence refers to the gradually occurring, slow-moving destruction caused by deforestation, climate change, toxic chemicals spreading in ecosystems, and so on.

In debating whether the Anthropocene calls us to rethink the "role of traditional scholarly activities" and usher in a "new Earth politics,"[15] we would examine why futures thinking has been ruled by popular books, articles, and TED Talks from science journalists, science fiction authors, Silicon Valley insiders, *Wired* magazine devotees, "transhumanist" human enhancement tech enthusiasts, life extension aficionados, and environmental activists. We would ask: why let these familiar futurologists dominate the conversation? After all, hundreds of geotechnical engineers, repository safety modelers, geologists, and nuclear waste regulators out there have already, for decades, been reckoning futures near and deep in just as, or perhaps more, rigorous ways. Why not add greater disciplinary, cultural, racial, gender, age, and class diversity to the array of voices speaking for the future today? Why not get more social scientists and humanities scholars involved?

After years of education in long-term thinking, Musk's one-hundred-thousand-year question would not only seem a lot less wacky and aloof; it would also seem more accessible. We would be equipped with far more ideas, information, frameworks, concepts, methods, principles, and vocabularies to draw on when answering it. This would help us avoid getting mired in the bipolar extremes of "horror and hope, nightmare and dream, destruction and creation, dystopia and utopia," which often accompany futures thinking.[16] It would help us feel less paralyzed by overwhelming Anthropocene pessimism and feel less "abstracted out of significance" by deep time's harrowing breadth.[17] It would provide counterpoints to the sensationalized deep time aesthetics of sublime awe, terror, mystery, horror, profoundness, and unsurmountable complexity described in chapter 2. It would provide alternatives to the feelings of existential anxiety, cosmic loneliness, or meaninglessness they can evoke.

It would train us in the "art of noticing,"[18] helping us reveal short-termist blind spots in our thinking that might otherwise go unnoticed.

All of this would prepare us for our final capstone essay assignment. Before graduating high school, each student would need to answer the question: "How can I be a good ancestor?"[19] With long-termist learning as our common educational core, we would feel more open to engaging with the complex, uncertain tomorrows before us. From there, we could begin building more long-sighted worlds.

LONG-SIGHTED WORLDS

Let's continue our thought experiment. Let's imagine we live in a society very different from our own. We recently completed our formal education. We live in a city dotted with signs and plaques describing how local landscapes appeared at different points in geological history. Their inscriptions bring features of past glaciers, asteroid collisions, flora and fauna, volcanic activity, and waterways into our everyday awareness. Smartphone map applications not only help us navigate what regions look like today, but also depict what regions looked like decades, centuries, and millennia ago. We read news articles about popular augmented reality applications that, with the help of a special headset, immerse us virtually in our region's distant past.

In this alternative society, a program called Future Sister Cities, now widely popular, has been inspired by the CGIAR climate analogues project described in chapter 1. Through it, thousands of towns, cities, and villages set up knowledge-sharing partnerships with counterparts elsewhere in the world—with communities harboring ecosystems resembling those that their own town, city, or village is forecasted to harbor after decades of climate change. Equally popular is a program called Past Sister Cities, inspired by Safety Case experts' natural analogue studies. Through it, thousands of towns, cities, and villages partner with still other far-off communities—those home to ecosystems similar to those alive in their town, city, or village's very distant past. The communities present local artifacts to one another as gifts. The artifacts become Analogue Monuments in one another's public squares. The monuments help locals achieve distance from their present-day surroundings. All these tools provide us with, to use chapter 1's terms, interscalar vehicles for moving, intellectually, across time and place.[20] They help us see "every outcrop" as a "portal to an earlier world," or to a coming one.[21]

Catering to publics with a wide-ranging time-literacy education, popular podcasts and radio talk shows host lively discussions about time. They consider philosopher Edmund Husserl's notion of internal time consciousness,[22] philosopher Henri Bergson's theories of time duration,[23] Albert Einstein's understanding of space-time, and long-sighted books like anthropologist David Graeber's *Debt: The First 5,000 Years*.[24] They discuss the *The Reckoning of Time*, a treatise written by the eighth century English Benedictine monk Bede, which examined the ancient calendrical systems of the Romans, Anglo-Saxons, Greeks, Hebrews, and Egyptians. J. G. Ballard's 1962 science fiction novel *The Drowned World* inspires a blockbuster film widely seen as foretelling climate change's looming perils.[25] The film is about a globally warmed Earth struggling to adapt to flooding cities, melting polar ice caps, thawing tundra, and rising sea levels.

On the internet, netizens square off in public contests in which competitors propose, debate, and scrutinize propositions about the future. The competitions are set up similar to debate team matches. Often, these events are organized around the simple logical patterns such as if/then or input/output—or other patterns we drew on to speculate about futures in chapter 2. On television gameshows, contestants face off in tournaments modeled on social scientists Ranjit Singh and Joan Donovan's "The Future Will Be Terrible" exercise: they envision "techno-dystopias" by taking two existing technologies and imagining how they could be combined in insidious ways. They then try to "sell" their sinister innovations as profitable to a mock panel of angel investors.[26] This ironic show not only entertains audiences who see it as a matter of pride to be a sophisticated futures thinker—much like it is to be a skilled orator, a quick wit, or a trivia whiz. It also raises awareness about future threats, and helps officials working for the Secretary for the Future develop new policies. (The Secretary for the Future is a presidential cabinet position Kurt Vonnegut proposed shortly before his death in 2007.)

Corporations, agencies, and institutes with long-term impacts often host the multitimescale awareness training courses, boot camps, and employee retreats proposed in chapter 3. They house chapter 3's long-term safety departments too: a days division, a decades division, a centuries division, and a multimillennial division. To help employees achieve distance from their work, they set aside quiet meditation rooms for doing introspective thought experiments in tacking back-and-forth across different timescales and other perspectives. The multimillennial division practices self-critique similar to the biosphere experts' knowledge quality assessment work described in chapter 1. It has, on its walls, motivational posters displaying its mantra: "Accept Futility, Reckon Deep Time Anyway." Other posters, at odds with Seppo's screensaver, read: "More Deep Time Research Is Needed." The division develops data-driven ways of weaving models and scenarios together into complex depictions of future worlds. They draw on number crunching, information collection, and computer-simulation techniques like those the Safety Case experts used in chapter 2. However, following the "multiple lines of reasoning" principle, they also develop detail-driven qualitative scenarios, analogue arguments, and more freewheeling speculations. Their work routinely captures the imagination of society as a whole, which tends to associate high intelligence with one's ability to toggle back and forth or zoom in and out on different timespans, much as we did in chapter 3.

Meanwhile, chapter 3's Global Deep Time Reckoning Association unites thousands of long-sighted experts from different fields into a vibrant community with vast influence. Membership is seen as a prestigious honor. Members envision future worlds from several different scales, levels, and frames of analysis, as they socialize and tinker with long-term ideas in its Perspective-Exchanging Parlors. They discuss how ancient bacteria has reawakened because of rising global temperatures, after being frozen in Siberian soil for hundreds of thousands of years; they debate the merits of the RAND Corporation's long-term probability forecasts and strategic planning methods, which the US military has used since the 1950s. Arguing over whether postmodern cynicism has fostered widespread distrust in human hearts' and minds' capacities for achieving a more perfect long-termism, they ask whether left-leaning activists and critical scholars have grown too pessimistic about technology, melancholic about environmental futures, or shy about answering grand theoretical questions. Scrutinizing Christian concepts of time such as eternity, everlastingness, sempiternity, and foreverness, they reflect on theological concepts such as the *nunc aeternum* (the "eternal now")[27] or the *pro chronon aionion* (the Bible's time before time started).[28] They consider the Long Now Foundation's proposal that, to help us think in ten-thousand-year timespans, we should express, say, the year 2020 instead as "02020." Some ask whether "00002020" would be more appropriate.

The members of the Global Deep Time Reckoning Association develop a policy brief on whether the fossil fuel, nuclear, or plastics industries should require special training and certification in long-term planning, ethics, and stewardship. Some suggest that this should apply not only to the industries' experts and CEOs, but also to their assistants, administrators, accountants, IT personnel, lawyers, and others across the organization. Before working in the industry, they argue, all must first be made aware of the long-term costs versus benefits, to humanity and to the planet, of doing so.

Corporations, agencies, and institutes that employ deep time reckoners document, catalog, and disseminate their knowledge so it does not vanish when they die. Pressured by publics who support long-termism, they afford deep time reckoners special privileges and responsibilities. As proposed in chapter 4, they have deputies and protégés on standby as replacements. When they travel, they do not share the same vehicles. Their group work is distributed across different office locations. Successors take seriously their duties to inherit and refine their predecessors' long-termism.

Anthropologists, hired as consultants, interview deep time reckoners and their colleagues to help back-up their thinking in databases. Each deep time reckoner's most important findings, forecasts, scenarios, ideas, models, interview recordings, meetings transcripts, and notebooks get backed up in several archives in several vaults across the world. To reduce digital obsolescence risks, archivists and librarians store key information on "Rosetta Disks": glass and steel orbs with over 26,000 pages of text carved into them, which they hope will be readable by any future society with technology that can magnify the words to one thousand times their size.[29] The idea is that societies of tomorrow will be more likely to have simple microscope-like technology than any specific digital storage hardware or software of today. These sturdy, fire-resistant orbs are seen as a more durable form of long-term information storage than paper as well.

UNESCO oversees the Global Deep Time Reckoning Information Repository. Concerned citizens regularly collect expert-vetted information from its archives to sharpen their sophistication in doing chapter 1 and 2's futurological mental workouts. They also learn from the Global Deep Time Reckoning Association's policy proposals, publications, think-pieces, and YouTube videos. Bloggers, media pundits, government staffers, and academics draw on these sources to make long-termist arguments, while billions hope that we, over the next few decades, will slowly inch toward accomplishing our shared mission of fending off Anthropocene planetary collapse.

BUILD IT!

We do not, of course, live in these imagined worlds. In this sense, they are unreal—merely fictions. However, our capacities to envision alternative futures, and to work toward building them, are very real. Depictions of utopian or dystopian tomorrows, even those millennia from now, can have powerful, concrete effects on the world today. Engaging with them can help us build more time-literate societies. That is why futurological thought experiments like these are not playful games; they are serious acts of intellectual problem-solving. It is why the Safety Case's analogue studies of far futures are uniquely valuable, even though they are, at the end of the day, mere approximations. It is why Safety Case models can be enlightening even if they never offer perfect representations of what will definitely happen in the future. It is why, in each chapter, I have invited anyone who picks up this book to join me and my informants in reckoning futures in all sorts of ways. After all, the reckonings of *Deep Time Reckoning* were not final verdicts; they were open-ended brainstorming pathways handed off to fellow denizens of the Anthropocene who will, I hope, continue on in their own long-termist journeys—even after putting this book down.

Luckily for us, our world is already home to many different inspirational expert tools for widening the timescales of our thinking. Most are overlooked by the public at large. My examples were drawn from the deep time reckoning Safety Case experts of Finland's nuclear waste repository. But many others out there are just waiting to be popularized, repurposed, and mobilized to help dial back some of our most self-destructive short-sighted tendencies. Librarians and library science educators are, for instance, increasingly developing long-term information storage projects in response to today's fast-changing publishing landscape. What can we learn from their efforts? What can we learn from the long-sighted work of climate scientists, reinsurance industry professionals, archaeologists, strategic planners, urban infrastructure designers, or existential risk scholars? Quite a lot, I reckon. And this is grounds for optimism.

Yet our optimism must remain guarded and measured. Expertise is not dead, but it has been deflated. Experts' futurological methods are more sophisticated than ever. Long-sighted curricula, institutions, and societies are more necessary than ever, as new barriers to long-termism have taken root. Countless experts—historians, astrophysicists, geophysicists, nuclear regulators, anthropologists, evolutionary biologists, archaeologists, and others—face political climates highly skeptical of expert knowledge. If their repertoires of long-termist techniques were to win more hearts and minds, the deflation of expertise could begin to reverse, despite the obstacles that remain.

One challenge, for experts and laypeople alike, is to find ways of recasting experts' long-sighted techniques in more publicly accessible, attention-grabbing, exciting formats. These formats must avoid alienating the public with too much jargon, while still doing justice to expert techniques' complexity and rigor. A second challenge, in cosmologist Martin Rees's words, is to counter the common "misperception" that there is something inherently "elite" about scientific thought. We need fresh formats in which sophisticated science can be "accessed and enjoyed" by all in "nontechnical words and images."[30] A third challenge is to show the public that "soft" disciplines like anthropology, philosophy, and history are equally as important for ensuring our long-term survival as "hard" disciplines like physics, engineering, or mathematics. A fourth challenge lies in overcoming biases against experts who boldly embark on Promethean endeavors to reckon deep time. It would be naïve for us to think that these experts' forecasts could ever be totally accurate (or even mostly accurate). However, it would also be naïve to reject the possibility that experts can progress, over time, in developing more robust, credible, nuanced methods of future-gazing.

My strategy has been to combine the methods of academic Anthropology, the long-termism of Finland's nuclear waste expertise, and the writing style of popular science journalism. What emerged is a format that resists both Anthropocene ecological degeneration and the intellectual downturn of the deflation of expertise. Taking a full twelve years to develop, my anthropological exploration was out of sync with today's twenty-four-hour news cycles' nonstop feeds of shocking stories. It plodded on slowly and carefully amid online media's whirlwind of information and disinformation, hysterical speculations, and fractious debates. Yet this book constitutes just one form of intellectual resistance. There are many other ways of countering shallow time discipline already out there today, just waiting to be discovered. If we all find our own, then real progress is possible.

A DEEP TIME RECKONING LEXICON

Adventurous learning A thoughtful intellectual approach of actively seeking out scientifically informed knowledge while embracing an anthropologically informed openness to careful, skeptical, critical listening to all sorts of diverse perspectives.

Anthropocene A name proposed for a troubled epoch in Earth's geological history ushered in by human transformations of our planet's climate, erosion patterns, extinctions, atmosphere, rock record, and more. Many of the proposed epoch's advocates argue that humans have become telluric forces of nature: agents not just of ecological but also geological change.

Archaeological Analogues Present-day artifacts, ruins, or remnants of past human activities that nuclear waste researchers study in order to make extrapolations, inferences, or projections about future nuclear waste repositories' engineered components or natural surroundings.

Asiat riitelevät, eivät ihmiset A Finnish phrase meaning "the issues fight, not the people," which can serve as a model for thoughtful, civil, respectful debate in times of social, political, and ecological tumult.

Attunement Legal scholar Frank Pasquale's term for a principled embrace of receptivity, sensitivity, and appreciation of one's wider world. To embrace attunement is to resist yearning for mastery, escapism, or mechanistic thinking.

Big History An interdisciplinary academic field that studies history in the widest of timescales—far beyond social or historical time horizons—from the Big Bang to the present.

Biosphere Assessment A subsection of Posiva's nuclear waste repository Safety Case portfolio, which examines the paths of radionuclides that could, in worst-case scenarios, escape from the Olkiluoto repository and be released into Western Finland's landscapes. Biosphere models extrapolate about how radionuclides might travel around in distant future lakes, rivers, forests, fields, and bogs. The biosphere report explores questions like: At what pace will Finland's shoreline continue expanding outward into the Baltic Sea? What happens if forest fires, soil erosion, or floods occur? How and where will lakes, rivers and forests sprout up, shrink and grow? What role will climate change play in all this?

CGIAR Climate Analogues Project CGIAR's Research Program on Climate Change, Agriculture, and Food Security's online tool for exploring rainfall and climate forecasts in various future locales across the world. This tool makes analogies comparing these future regions' conditions with various present-day regions, which are already seeing similar conditions.

Complementary Considerations A subsection of Posiva's nuclear waste repository Safety Case portfolio, which contains a hodgepodge of public relations information and qualitative evidence designed to persuade wider audiences of the repository's strengths. The report contains most of the Safety Case experts' analogue research findings and is seen as filling knowledge gaps that computer modeling and engineering calculations alone cannot fill.

Deflation of expertise A term proposed for our current historical moment, in which political power is commonly gained through populist mockery of expert authority, when experts' voices are drowned out by the noisy clamors of knee-jerk tweets and self-published blogs, and when experts' inquisitive spirits are dulled by new corporate-managerial reforms and constrained by bureaucratic protocols. In this context, it is increasingly difficult to put forth bold, evidence-driven, intellectually ambitious visions of the Earth's future. This situation is exacerbated by widespread skepticisms of technocratic knowledge, liberal arts education, scientific research on the environment, and even the very possibility of there being verifiable facts, truth, or a single shared reality.

Existential risk A severe future risk that could destroy global human flourishing, cripple civilizational progress, trigger human extinction, or end life on Earth. Examples include pandemic disease, a supervolcano eruption, a total nuclear war, a large asteroid impact, or a hostile insurrection of artificially intelligent machines.

Geoengineering A term for proposals to technologically manipulate our planetary environment in order to strategically reduce the effects of climate change. Examples include proposals to pour fertilizers into the Earth's oceans to raise their carbon dioxide uptake, or to pump reflective particles into the Earth's atmosphere to deflect sunlight.

Global Deep Time Reckoning Association A proposed international organization that aims to unite long-sighted experts of many kinds—astrophysicists, geologists, historians, cosmologists, anthropologists, evolutionary biologists, archivists, paleontologists, nuclear waste scientists, climatologists, philosophers, archaeologists, and other long-term thinkers—into self-identifying as a pragmatic interdisciplinary community that today's societies must rely on to avert planetary collapse.

Global Deep Time Reckoning Information Repository A proposed international archive in which prolific deep time reckoning experts' findings, forecasts, scenarios, ideas, models, interview recordings, meetings transcripts, and notebooks could be archived, conserved for posterity, and disseminated more widely across populations, educational institutions, and media outlets.

Hiljainen tieto The Finnish word for Michael Polanyi's concept of experience-based, "tacit knowledge" that, in being deeply inscribed in one's intuitions or muscle memory, cannot be fully recorded in technical manuals or backed up in databases. This form of knowledge is often contrasted with explicit, formal, or codified knowledges, which can be more easily transmitted via text, how-to guides, or digital media.

Hiljaisuus A Finnish word for "quietude" or "silence," often associated with being *omissa oloissaan* ("to oneself") in inward contemplation. When in Finland, one can embrace *hiljaisuus* in Helsinki's Kamppi Chapel, also known as the "Chapel of Silence": an ecumenical space where visitors from any religious tradition or life philosophy can meditate quietly in a serene wooden structure in one of Finland's busiest areas. Embracing contemplative self-reflection can help us cultivate the thoughtfulness necessary for transcending today's intellectual and environmental crises.

Infinition Philosopher Emmanuel Levinas's term for a complexity "overflowing the thought that thinks it," leading to the "overflowing of the idea by its ideatum." Gazing into distant futures or pasts can have this effect. Since there is always more of deep time to know, its complexities can never be fully lassoed into place.

Interscalar vehicles Historian Gabrielle Hecht's term for objects that can draw scholars and others to "move simultaneously through deep time and human time, through geological space and political space." Studying them can provide "means of connecting stories and scales usually kept apart."

Iteration Finland's nuclear waste repository Safety Case experts' stepwise approach of developing progressively new and improved versions, or "iterations," of their deep time reckoning models before each successive repository license application submission deadline. Between the publication of each new version of the Safety Case, lessons learned from the previous iteration can be factored in to improve future iterations. Updated iterations are to be released until the Olkiluoto repository's planned decommissioning around 2120.

KBS-3 A spent nuclear fuel repository design concept used in Sweden and Finland. Posiva's KBS-3 repository plan involves sliding spent nuclear fuel bundles into tube-shaped cast iron insert containers. They then plan to slide the inserts inside large copper canisters at an encapsulation plant in Olkiluoto, now under construction. Once emplaced underground in one of the repository's deposition holes 400–450 meters underground, the canisters are to be surrounded by an absorbent bentonite clay. The clay is to serve as a "buffer" between the canisters and the bedrock. Posiva's KBS-3 design has mostly been derived from the Swedish nuclear waste company SKB's research and development work.

Knowledge Quality Assessment A self-critical section in Posiva's Olkiluoto nuclear waste repository Safety Case's biosphere modeling report, which aimed to increase confidence in the experts' far future forecasts. Its approach was to openly admit to—and then systematically detail—various uncertainties and reductive assumptions found in the biosphere modelers' own knowledge base.

Lake Lappajärvi A crater lake in Southern Ostrobothnia, Finland, that formed over 73 million years ago after a meteor crashed into the Earth. Among Finland's nuclear waste repository Safety Case experts, the crater became an analogue for how the country's landscape may or may not change over the next several ice ages.

Long Now Foundation Established in San Francisco in 1996, the Long Now Foundation aims to "provide a counterpoint to today's accelerating culture," to "help make long-term thinking more common," and to "foster responsibility in the framework of the next 10,000 years." This organization is known for its plans to build a clock, hundreds of feet tall, which will tick continuously for ten millennia. Built as a monument to long-term thinking, the Clock of the Long Now is sited deep in a West Texas mountain.

Multiple lines of reasoning Finland's nuclear waste repository Safety Case experts' strategy of intentionally having multiple teams of experts—each with different disciplinary backgrounds and intellectual tendencies—working in parallel on the same long-sighted challenges. To this end, some experts developed quantitative models, others developed natural analogue studies, others conducted mechanical tests on KBS-3 repository components, and so on. In Posiva's Safety Case portfolio, these arguments for repository performance were presented in parallel—offering a more robust and holistic set of arguments for repository safety.

Multiscalar perspective A holistic form of analysis that intentionally tacks back and forth between several different scales of generality and particularity—approaching a topic from multiple angles, levels, and perspectives. Such analyses gesture to how any given event, entity, or vision of the future can be viewed in several different ways, depending on the perspective from which one approaches it.

Multitemporal perspective A holistic form of analysis that intentionally approaches a topic from the perspective of a variety of different timespans. What emerges is a more dynamic and multidimensional understanding of a given event, entity, or vision of the future's causes and effects across futures and pasts. Multitemporal sensibilities can be cultivated in contemporary organizations by (a) instituting chapter 3's proposed workplace multitimescale awareness training workshops, retreats, or training sessions or (b) establishing chapter 3's proposed timescale-specific days divisions, decades divisions, centuries divisions, and multimillennial divisions.

Natural analogues Examples of present-day ecological or geophysical features, processes, or locales that nuclear waste experts study to make extrapolations, inferences, or projections about repositories' engineered components or natural surroundings in the future. Natural analogues research is compiled globally by the international Natural Analogue Working Group (NAWG) organization.

Offentlighetsprincipen Sweden's constitution's "Principle of Access to Official Information," which is aimed at fostering transparency among government organizations.

Oklo A uranium deposit in Gabon, Africa, that experienced self-sustaining, natural fission chain reactions roughly 1.7 billion years ago. Today, nuclear waste programs point to the radionuclides the fossilized reactor left behind as analogues for understanding how radionuclides released from today's nuclear waste repositories may or may not disperse in far futures.

Paskaporukka A playful nickname, translated from Finnish as "Shit Gang," for Finland's founding group of geologic disposal experts from the 1970s and 1980s.

Performance Assessment A section of Finland's Olkiluoto nuclear waste repository Safety Case, which consists of multiple quantitative models. The Performance Assessment was Posiva's primary systematic analysis of how the repository's mechanical parts, heat levels, nearby groundwaters, and so on may interact over the coming hundreds of thousands of years.

Pessimisti ei pety A Finnish phrase, translated as "a pessimist is not disappointed," which evokes a guarded approach to future events.

Pilkunnussijat A pejorative term, translated from Finnish as "comma-fuckers," used as an epithet for people who are too detail-obsessed for their own good.

Posiva Finland's nuclear waste management company, which is jointly owned by Finnish power companies Teollisuuden Voima Oyj and Fortum Power & Heat. Posiva is developing what will, in the 2020s, likely become the world's first operating underground disposal repository for high-level nuclear waste. This facility is currently being built in Western Finland's granite bedrock, deep below the island of Olkiluoto in the Gulf of Bothnia.

Predecessor preservation An ethic of motivating younger generations to work toward carefully absorbing, tending to, and disseminating insights from prolific deep time reckoners so that their contributions to humanity's long-termism do not die off upon their biological deaths.

Radionuclide Transport A subsection of Posiva's nuclear waste repository Safety Case portfolio, which models the possible future routes of radionuclides that could, in unlucky conditions, escape from Finland's Olkiluoto repository and travel through groundwater channels toward Western Finland's surface.

Reckonings A name I've given to this book's open-ended guidance for embracing long-term thinking. These reckonings, found at the conclusions of each chapter, are not meant as hard instructions telling anyone how to think or act. They are suggested starting points for orienting others in embarking on long-termist learning-journeys of their own.

Ressentiment A French word for the psychological state of being weighed down by repressed emotions of envy or anger at the past, which the sufferer can no longer act on. This often results in self-deprecation or self-abasement. Philosopher Friedrich Nietzsche saw *ressentiment* as a defining feature of Judeo-Christian "slave morality."

Safety Case A safety assessment report that provides evidence-based arguments for a given technology's or facility's future health or environmental acceptability. Often mandated for submission to government regulatory agencies, safety cases are developed for new aviation technologies, medical devices, or railway infrastructures. Nuclear waste repository licensing safety cases, like the one Posiva developed for submission to Finland's nuclear regulator STUK, often make predictions about geological, hydrological, and ecological conditions tens or hundreds of millennia in the future.

Säteilyturvakeskus (STUK) Finland's Radiation and Nuclear Safety Authority is the country's nuclear regulator and radiation monitoring agency. In 2015, STUK, in conjunction with Finland's Ministry of Economic Affairs and Employment, approved Posiva's Safety Case's projections of conditions the Olkiluoto nuclear waste repository may face over the coming tens of thousands, hundreds of thousands, or even millions of years.

Shallow time discipline The powerful cultural, economic, political, and technological forces that interact to fix our attentions on the radical short-term. Examples include twenty-four-hour news cycles, the boom-bust-or-buyout tempos of startup companies, throwaway consumerism, lightning-fast financial transactions, short-lived political electoral terms, digital marketplace tempos, contingent employment policies, fickle popular culture trends, corporate fixations on quarterly earnings, and more.

Slow violence A concept, popularized by English professor Rob Nixon, referring to the gradually occurring, slow-moving destruction caused by deforestation, climate change, toxic chemicals spreading in ecosystems, and other more gradually unfolding crises. Instances of slow violence often unfortunately fail to win hearts and minds or incite political action in today's spectacle-driven media landscape.

Successor stewardship A proposed code of ethics for deep time reckoning experts, which focuses on sustaining and refining long-term thinking patterns across the generations. This means that each new generation must see themselves as both (a) mouthpieces for transmitting predecessors' long-term thinking patterns and (b) reformers working to end the transmission of impediments to long-termism. Successor stewardship involves responsibly selecting which aspects of a predecessor should be mimicked and replicated versus which should be discouraged and discarded by new generations.

Suuret nälkävuodet The "great hunger years" famine that killed almost 9 percent of Finland's population in just three years, from 1855 to 1868.

Svalbard Global Seed Vault A program to back up plant seeds in an underground Arctic repository to ensure against genetic diversity loss during future global crises. This gene-bank vault is located on Spitsbergen Island in Norway's Svalbard archipelago.

Timefulness A concept popularized by geologist Marcia Bjornerud, which reminds us to be "mindful that this world contains so many earlier ones, all still with us in some way—in the rocks beneath our feet, in the air we breathe, in every cell of our body." Timefulness is one route to greater time-literacy.

Time reckoning A social scientific concept for how different individuals or communities understand, represent, measure, or grapple with various spans of time, and their past and future horizons. Studying how people approach time's intervals, scales, horizons, and durations can reveal how they see themselves and their worlds.

Timescapes A sociological concept, developed by Barbara Adam, that attends to how different paces, scales, durations, sequences, and modalities of past, present and future converge to form context-specific "scapes" of time. For example, tempos of democratic electoral schedules, tempos of capitalist commodity production, and tempos of communication and transportation speeds may converge, in unique ways, to stoke environmental problems.

Tyhjän puhuminen A Finnish word for "empty talk," which has negative connotations. Many (but not all) Finns are seen as disliking unnecessary small-talk, long-winded tangents, and unconcise speech—embracing straightforward, less verbose modes of self-expression.

Waste Isolation Pilot Plant (WIPP) The United States' transuranic defense nuclear waste repository, located deep in a salt mine near Carlsbad, New Mexico. In the late 1980s and early 1990s, the US Department of Energy assembled interdisciplinary teams to help brainstorm ideas for warning monuments to deter far future tomb-raiding treasure hunters, archaeologists, local communities, and miners from digging near the repository.

Yucca Mountain License Application A US Department of Energy facility performance assessment document developed to seek US Nuclear Regulatory Commission approval for the country's now-defunct high-level nuclear waste repository program in Yucca Mountain, Nevada. The License Application includes Total System Performance Assessment models that gaze one million years into the future.

NOTES

INTRODUCTION

1. David Armitage and Jo Guldi, "Bonfire of the Humanities," *Aeon*, October 2, 2014, http://aeon.co/magazine/society/how-history-forgot-its-role-in-public-debate/.

2. See R. Monastersky, "Anthropocene: The Human Age," *Nature* 519, no. 7542 (2015): 144; see also W. Steffen, J. Grinevald, P. Crutzen, and J. McNeill, "The Anthropocene: Conceptual and Historical Perspectives," *Philosophical Transactions: Mathematical, Physical and Engineering Sciences* 369, no. 1938 (2011): 842–867.

3. Richard Irvine, "Deep Time: An Anthropological Problem," *Social Anthropology* 22, no. 2 (2014).

4. Stewart Brand, *The Clock of the Long Now: Time and Responsibility* (New York: Basic Books, 1999).

5. Martin Rees, *On the Future: Prospects for Humanity* (Princeton, NJ: Princeton University Press, 2018), 3, 217.

6. Tim Ingold, *Evolution and Social Life* (Cambridge: Cambridge University Press, 1986), 129.

7. Michael Welker and John Polkinghorne, *The End of the World and the Ends of God: Science and Theology on Eschatology* (Trinity: Harrisburg, 2000), 8.

8. Gregory Benford, *Deep Time: How Humanity Communicates across Millennia* (New York: Avon, 2000); Stephen Jay Gould, *Time's Arrow, Time's Cycle: Myth and Metaphor in the Discovery of Geological Time* (Cambridge, MA: Harvard University Press, 1987); Robert Macfarlane, *Underland: A Deep Time Journey* (New York: W. W. Norton, 2019); Martin Rudwick, *Scenes from Deep Time: Early Pictorial Representations of the Prehistoric World* (Chicago: University of Chicago Press, 1992); Daniel Lord Smail, *On Deep History and the Brain* (Berkeley: University of California Press, 2008).

9. E. E. Evans-Pritchard, "Nuer Time-Reckoning," *Africa* 12 (1939).

10. Ernest Gellner, *Thought and Change* (London: Weidenfeld and Nicolson, 1964), 1.

11. If their livelihoods were improved, would billions of new long-term thinkers come to the fore? Or would their attentions merely refixate on another set of short-terms—those that distract the currently better-off from long-sightedness?

12. United Nations, "World Population Projected to Reach 9.8 billion in 2050, and 11.2 Billion in 2100," Department of Economic and Social Affairs (2017), https://www.un.org/development/desa/en/news/population/world-population-prospects-2017.html.

13. United Nations, "68% of the World Population Projected to Live in Urban areas by 2050," Department of Economic and Social Affairs (2018), https://www.un.org/development/desa/en/news/population/2018-revision-of-world-urbanization-prospects.html.

14. Annelise Riles, "The Politics of Expertise in Transnational Economic Governance: Breaking the Cycle," *Social Sciences of Crisis Thinking* 10, no. 3 (2017).

15. Annelise Riles, *Financial Citizenship: Experts, Publics and the Politics of Central Banking* (Ithaca, New York: Cornell University Press, 2018), 2.

16. Meridian 180, "Forum Summary—What Role for Global Intellectuals?" (2011), https://meridian-180.org/en/forums/forum-summary-what-role-global-intellectuals.

17. Sometimes I engaged with Nordic nuclear experts in a more participatory spirit, too. In February 2012, for example, I gave a presentation for Posiva's management group. In summer 2013, I spent two days at a Safety Case expert's family's *kesämökki* ("summer cottage") and then we visited Finland's Tytyri limestone mine. We visited a uranium and copper deposit geological research site together. Another informant brought me to a laboratory working on the welding and nondestructive testing of the repository's copper nuclear waste canisters. In June 2013, I attended the European Nuclear Society's Young Generation Forum event in Stockholm. While in Sweden, I visited Swedish nuclear waste disposal company SKB's underground Äspö Hard Rock Laboratory—a facility similar to Posiva's Onkalo underground laboratory—and its CLAB interim storage facility for spent nuclear fuel. In September 2013, I attended the World Nuclear Association's annual symposium in London.

18. To break my field materials down further: I recorded (a) fifty conversations with engineers, modelers, geologists, and other experts who could be considered insiders to Posiva's repository project; (b) ten conversations with experts who could be considered insiders to STUK's regulatory oversight of Posiva's repository project; (c) thirteen conversations with artists, NGO workers, environmentalists, and activists who were skeptical of Posiva's work; (d) eleven conversations with nuclear sector insiders who worked on nuclear energy production (as opposed to nuclear waste disposal); (e) eleven conversations with critical experts (in, e.g., geology, engineering, social sciences, and architecture) skeptical of Posiva's work; (f) ten conversations with experts in academia and/or the private sector who specialized in the financial-legal aspects of nuclear energy in Finland; (g) six conversations with political actors who had views on Finland's nuclear energy/waste projects; (h) six conversations with

bloggers, industry lobbyists, and environmental activists who held broadly pro-nuclear and/or "ecomodernist" views; and (i) four conversations with miscellaneous experts, artists, and/or members of the public who had noteworthy perspectives on nuclear energy/waste or culture in Finland.

19. Finland's Ministry of Economic Affairs and Employment, "Spent Fuel Disposed of in Finland," https://tem.fi/en/spent-nuclear-fuel (accessed February 24, 2019).

20. The Anthropocene concept has many precedents. In the late eighteenth century, France's Comte de Buffon explained how the "entire face of the Earth today bears the imprint of human power" (C. Bonneuil and J.-B. Fressoz, *The Shock of the Anthropocene: The Earth, History, and Us* [London: Verso, 2016], 4). Around 1873, Italian priest and geologist Antonio Stoppani proposed the "Anthropozoic Era," an idea taken up in American diplomat and philologist George Marsh's 1874 *The Earth as Modified by Human Action*. In 1922, British geologist Robert Sherlock published *Man as a Geological Agent: An Account of His Actions on Inanimate Nature*. Soviet earth scientist Vladimir Vernadsky, a father of the concept of the Earth's biosphere, wrote of a "psychozoic era" marked by the "influence of consciousness and collective human reason upon geochemical processes" in his 1927 essay collection on geochemistry. Fairfield Osborn's 1948 *Our Plundered Planet* introduced the idea of "Man: A New Geological Force." A 1955 Princeton event titled "Man's Role in Changing the Face of the Earth" revived these themes. In the decades to follow, ideas of humanity as a geological force interspersed the development of what would become Earth systems science, but the topic remained something of a sideline item (Steffen et al., "The Anthropocene," 843–844). Biologist Eugene Stoermer, however, has used the term Anthropocene informally since at least the 1980s. Andrew Revkin, in a 1992 book on global warming, speculated that future earth scientists would give the name Anthrocene to a post-Holocene "geological age of our own making" (Revkin, *Global Warming: Understanding the Forecast* [New York: Abbeville Press, 1992], 55).

21. Jan Zalasiewicz, "Response to Adrian J. Ivakhiv's 'Against the Anthropocene' Blog Post," *Immanence* 7 (July 2014).

22. Steffen et al., "The Anthropocene."

23. J. Zalasiewicz, C.N. Waters, M. Williams, A.D. Barnosky, A. Cearreta, P. Crutzen, E. Ellis, M.A. Ellis, I. J. Fairchild, J. Grinevald, R. Leinfelder, J. McNeill, C. Poirier, D. Richter, W. Steffen, D. Vidas, M. Wagreich, A. P. Wolfe, and A. Zhisheng, "When Did the Anthropocene Begin? A Mid-Twentieth-Century Boundary Level Is Stratigraphically Optimal," *Quaternary International* 383 (2015): 204–207.

24. Dipesh Chakrabarty, "The Climate of History: Four Theses," *Critical Inquiry* 35, no. 2 (2009): 197–222.

25. Morton's "agrilogistics" include simple algorithms that today's societies still perpetuate, like "eliminate contradiction and anomaly" and "establish boundaries between the human and the nonhuman." His argument is that they have, since the dawn of agriculture, gone "viral" across the globe—later requiring industrialization to persist. See Timothy Morton, "How I Learned to Stop Worrying and Love the Term Anthropocene," *Cambridge Journal of Postcolonial Literary Inquiry* 1, no. 2 (2014): 259.

26. Bruno Latour, "Telling Friends from Foes in the Time of the Anthropocene," in *The Anthropocene and the Global Environmental Crisis: Rethinking Modernity in a New Epoch*, ed. C. Hamilton, F. Gemenne, C. Bonneuil (London: Routledge, 2015), 146.

27. Bruno Latour, "Anthropology at the Time of the Anthropocene: A Personal View of What Is to Be Studied," 113th AAA Meeting Distinguished Lecture (Washington, DC, November 5, 2014), http://www.bruno-latour.fr/node/607.

28. Nikolas Rose, "The Human Sciences in a Biological Age," *Theory, Culture, & Society* 30, no. 1 (2013).

29. Rees, *On the Future*, 3.

30. Rudwick, *Scenes from Deep Time*, 2.

31. Julia Thomas, "History and Biology in the Anthropocene: Problems of Scale, Problems of Value," *American Historical Review* 119, no. 5 (2014).

32. Andreas Malm, "The Anthropocene Myth," *Jacobin*, March 30, 2015, https://www.jacobinmag.com/2015/03/anthropocene-capitalism-climate-change/.

33. Jason Moore, "The Capitalocene, Part I," *Journal of Peasant Studies* 44, no. 3 (2017).

34. Erik Swyngedouw, "Anthropocenic Promises," lecture at the CERI, Sciences Po Paris, France, June 2, 2014.

35. Donna Haraway, "Anthropocene, Capitalocene, Plantationocene, Chthulucene: Making Kin," *Environmental Humanities* 6 (2015): 159–165.

36. Bonneuil and Fressoz, *Shock of the Anthropocene*.

37. STUK, Disposal of Nuclear Waste Guide YVL D.5/15 (2014).

38. Anna Weichselbraun, "Not Talking about Disarmament at the IAEA," *Anthropology News*, July 19, 2018.

39. Michael Madsen, *Into Eternity: A Film for the Future* (Films Transit International, 2010).

40. Kristin Shrader-Frechette, *Burying Uncertainty: Risk and the Case against Geological Disposal of Nuclear Waste* (Berkeley: University of California Press, 1993); Kristin Shrader-Frechette, "Mortgaging the Future: Dumping Ethics with Nuclear Waste," *Science & Engineering Ethics* 11 (2005): 518–520.

41. Brian Bloomfield and Theo Vurdubakis, "The Secret of Yucca Mountain: Reflections on an Object in Extremis," *Society & Space* 23, no. 5 (2005).

42. Benford, *Deep Time*. For an anthropological perspective on WIPP, see Vincent Ialenti, "Waste Makes Haste: How a Campaign to Speed up Nuclear Waste Shipments Shut Down the WIPP Long-term Repository," *Bulletin of the Atomic Scientists* 74, no. 4 (2018).

43. Peter Galison and Robb Moss, directors, *Containment* (film, 2015).

44. Gabrielle Hecht, "Interscalar Vehicles for an African Anthropocene: On Waste, Temporality, and Violence," *Cultural Anthropology* 33, no. 1 (2018).

45. K. M. Trauth, S. C. Hora, and R. V. Guzowski, *Expert Judgment on Markers to Deter Inadvertent Human Intrusion into the Waste Isolation Pilot Plant*, Sandia National Laboratories (SAND-92-1382, 1993).

46. David Farrier, "Deep Time's Uncanny Future Is Full of Ghostly Human Traces," *Aeon*, October 31, 2016, https://aeon.co/ideas/deep-time-s-uncanny-future-is-full-of-ghostly-human-traces.

47. John Playfair, "Of Dr. James Hutton," in *The Works of John Playfair* (London: George Ramsay, 1822 [1805]).

48. Michael Tomko, "Varieties of Geological Experience: Religion, Body, and Spirit in Tennyson's In Memoriam and Lyell's Principles of Geology," *Victorian Poetry* 42, no. 2 (2004): 113–133.

49. Vincent Ialenti, "Adjudicating Deep Time: Revisiting the United States' High-Level Nuclear Waste Repository Project at Yucca Mountain," *Science & Technology Studies* 27, no. 2 (2014).

50. Isabelle Stengers, "Accepting the Reality of Gaia," in *The Anthropocene and the Global Environmental Crisis: Rethinking Modernity in a New Epoch*, ed. C. Hamilton, F. Gemenne, C. Bonneuil (London: Routledge, 2015).

51. Breakthrough Institute, "Breakthrough Dialogue 2015: The Good Anthropocene" (2015), https://thebreakthrough.org/articles/breakthrough-dialogue-2015.

52. Shawn Lawrence Otto, "Antiscience Beliefs Jeopardize U.S. Democracy," *Scientific American*, November 1, 2012, http://www.scientificamerican.com/article/antiscience-beliefs-jeopardize-us-democracy/.

53. Tom Nichols, "The Death of Expertise," *Tomnichols.net* (blog), December 11, 2013, https://thefederalist.com/2014/01/17/the-death-of-expertise/.

54. Decca Aitkenhead, "Peter Higgs: I Wouldn't Be Productive enough for Today's Academic System," *Guardian*, June 12, 2013, https://www.theguardian.com/science/2013/dec/06/peter-higgs-boson-academic-system.

55. Richard van Noorden, "Publishers Withdraw More than 120 Gibberish Papers," *Nature*, February 24, 2014, http://www.nature.com/news/publishers-withdraw-more-than-120-gibberish-papers-1.14763.

56. R. F. Harris, *Rigor Mortis: How Sloppy Science Creates Worthless Cures, Crushes Hope, and Wastes Billions* (New York: Basic Books, 2017).

57. David Graeber, "Are You in a BS Job? In Academe, You're Hardly Alone," *Chronicle of Higher Education*, May 6, 2018.

58. "Peak higher ed means we've reached the maximum size that colleges and universities can support. ... There's a qualitative aspect to all of this, namely that a lot of Americans think higher ed is in crisis. Increasing numbers of us are skeptical of its value, terrified of student loan debt, don't think college is needed for many jobs, etc. While anti-higher education feelings used to be a specialty of the political right, they've now crossed over into Democratic territory (cf. Paul Krugman's recent complaint, among many other examples). This is not an atmosphere likely

to send rising numbers of people to campus." Bryan Alexander, "Peak Education 2013," *BryanAlexander.org* (blog), September 18, 2013, https://bryanalexander.org /uncategorized/peak-education-2013/.

59. Mick Krever, "Has America Lost Its Ability to Dream Big?" *CNN.com*, February 11, 2014.

60. Juliet Eilperin and Brady Dennis, "Trump Proposes Change to Environmental Rules to Speed up Highway Projects, Pipelines and More," *Washington Post*, January 9, 2020.

61. Eurobarometer 62, *Public Opinion in the European Union: National Report Finland* (2004), http://ec.europa.eu/public_opinion/archives/eb/eb82/eb82_first_en.pdf.

62. Annukka Berg, "The Discursive Dimensions of a Decent Deal: How Nuclear Energy Evolved from Environmental Enemy to Climate Remedy in the Parliament of Finland," in *The Renewal of Nuclear Power in Finland*, ed. Matti Kojo and Tapio Litmanen (New York: Palgrave Macmillan, 2009), 97, 114.

63. Tapio Litmanen, "The Temporary Nature of Societal Risk Evaluation: Understanding the Finnish Nuclear Decisions," in *The Renewal of Nuclear Power in Finland*, 192, 198.

64. Matti Kojo, "The Revival of Nuclear Power in a Strong Administrative State," in *The Renewal of Nuclear Power in Finland*, 235.

65. Markku Lehtonen, "Reactions to Fukushima in Finland, France and the UK: Rupture or Continuity in the Nuclear Techno-Politics?" *An STS Forum on the East Japan Disaster* (2013), https://fukushimaforum.wordpress.com/.

66. Biljana Markova, "Finland Welcomes Disaster Risk Review," United Nations Office for Disaster Risk Reduction, July 14, 2014, http://www.unisdr.org/archive /38562.

67. Nic Newman, "Overview and Key Findings of the 2016 Report," Oxford's Reuters Institute (2016), http://www.digitalnewsreport.org/survey/2016/overview -key-findings-2016/.

68. Elina Kestilä-Kekkonen and Peter Söderlund, "Political Trust and Institutional Performance: Evidence from Finland 2004–2013," *Scandinavian Political Studies* 39, no. 2 (2016): 138–160.

69. Erik Kain, "The Finland Phenomenon: Inside the World's Most Surprising School System," *Forbes*, May 2, 2011, http://www.forbes.com/sites/erikkain/2011/05 /02/the-finland-phenomenon-inside-the-worlds-most-surprising-school-system/.

70. Laura Iisakka, ed., *Social Capital in Finland: Statistical Review* (Statistics Finland, 2006), https://www.stat.fi/tup/julkaisut/tiedostot/isbn_950-467-602-2_en.pdf.

71. Tomi Kankainen, "Voluntary Associations and Trust in Finland," *Research on Finnish Society* 2 (2009): 5–7.

72. Donal Carbaugh, Michael Berry, and Marjatta Nurmikari-Berry, "Coding Personhood Through Cultural Terms and Practices: Silence and Quietude as a Finnish 'Natural Way of Being,'" *Journal of Language and Social Psychology* 25, no. 3 (2006).

73. Evgenie Bogdanov, "The Principle of Trust," *Helsinki Times*, December 20, 2012, http://www.helsinkitimes.fi/columns/columns/expat-view/4825-the-principle-of -trust.html.

74. Rebecca Libermann, "Few Fukushima Fears for the Finns," *This Is Finland*, May 23, 2011.

75. Richard Black, "Nuclear Waste Plan Hangs on Trust," *BBC News*, September 21, 2010, http://www.bbc.com/news/science-environment-11378889.

76. Richard Black, "Finland Buries Its Nuclear Past," *BBC News*, April 27, 2006, http://news.bbc.co.uk/2/hi/science/nature/4948378.stm.

77. Martti Kalliala, Jenna Sutela and Tuomas Toivonen, *Solution 239–246 Finland: The Welfare Game* (Berlin: Sternberg Press, 2011).

78. Litmanen, "Temporary Nature of Societal Risk Evaluation"; Ari Lampinen, "An Analysis of the Justification Arguments in the Application for the New Nuclear Reactor in Finland," in *The Renewal of Nuclear Power in Finland*.

79. Case Pyhäjoki, "Artistic Reflections on Nuclear Influence," Call for Participants, Bioart Society, May 31, 2013, https://bioartsociety.fi/projects/case-pyhajoki/posts /call-case-pyhajoki-artistic-reflections-on-nuclear-influence.

80. Newman, *Digital News Report*.

81. Ray Kurzweil, *The Singularity Is Near: When Humans Transcend Biology* (New York: Viking Press, 2005).

82. Thomas Kuhn, *The Structure of Scientific Revolutions* (Chicago: University of Chicago Press, 1969).

83. Brand, *Clock of the Long Now*.

CHAPTER 1

1. Adrian Currie, *Rock, Bone, and Ruin: An Optimist's Guide to the Historical Sciences* (Cambridge, MA: MIT Press, 2018), 13.

2. Currie, *Rock, Bone, and Ruin*, 15.

3. T. Hjerpe, A. T. K. Ikonen, and R. Broed, "Biosphere Assessment Report 2009," Posiva Oy Databank (2009), 27, 141, http://www.posiva.fi/en/databank/biosphere_ assessment_report_2009.1867.xhtml#.VcPF6Ra8_dk.

4. Stefan Helmreich, "Extraterrestrial Relativism," *Anthropological Quarterly* 85, no. 4 (2012): 1125–1139.

5. Currie, *Rock, Bone, and Ruin*, 191.

6. Currie, *Rock, Bone, and Ruin*, 216.

7. At times, however, data derived from analogue sites got input into the Safety Case's models—which I explore in chapter 2.

8. Gabrielle Hecht, "Interscalar Vehicles for an African Anthropocene: On Waste, Temporality, and Violence," *Cultural Anthropology* 33, no. 1 (2018): 115, 134.

9. Hecht, "Interscalar Vehicles," 131.

10. Stefan Helmreich, "Waves: An Anthropology of Scientific Things," *Hau: Journal of Ethnographic Theory* 4, no. 3 (2014): 265–284.

11. Patricia Coates, "A Drive-Through Geology Lesson," *Washington Post*, November 18, 2005).

12. J. Zalasiewicz, C. N. Waters, M. Williams, A. D. Barnosky, A. Cearreta, P. Crutzen, et al., "When Did the Anthropocene Begin? A Mid-twentieth-century Boundary Level Is Stratigraphically Optimal," *Quaternary International* 383 (2015): 204–207.

13. The Conservation Fund, "Ice Age National Scenic Trail," Our Projects, https://www.conservationfund.org/projects/ice-age-national-scenic-trail (accessed December 31, 2019).

14. Preston Ni, "How to Unleash Your Creativity and Find Inspiration Today," *Psychology Today*, February 2, 2014.

15. Scott Kaufman, *Ungifted: Intelligence Redefined* (New York: Basic Books, 2013), 103.

16. Melanie Rudd, Kathleen Vohs, and Jennifer Aaker, "Awe Expands People's Perception of Time and Enhances Well-Being," *Psychological Science* 23, no. 10 (2012): 1130–1136.

17. Barbara King, "Atheists Feel Awe, Too," *National Public Radio—Cosmos & Culture Blog*, August 28, 2014.

18. Emmanuel Levinas, *Totality and Infinity: An Essay on Exteriority* (Pittsburgh: Duquesne University Press, 1969), 41.

19. Marilyn Strathern, *Partial Connections* (Savage, MD: Rowman & Littlefield, 1991).

20. Errol Morris, dir., *The Unknown Known: The Life and Times of Donald Rumsfeld* (film, Radius-TWC, 2013).

CHAPTER 2

1. G. N. Bailey, "Time Perspectivism: Origins and Consequences," in *Time and Archaeology: Time Perspectivism Revisited*, ed. S. Holdaway and L. Wandsnider (Salt Lake City: University of Utah Press, 2008): 21.

2. Hirokazu Miyazaki, "Saving TEPCO: Debt, Credit and the 'End' of Finance in Post-Fukushima Japan," in *Corporations and Citizenship*, ed. Greg Urban (Philadelphia: University of Pennsylvania Press), 130.

3. Here is a list of the kinds of reports found in the Safety Case portfolio: "Description of SNF [spent nuclear fuel]; description of EBS [the repository's engineered barrier systems]; performance assessment of repository's capacity to provide containment and isolation of SNF for as long as it remains hazardous; definition of the lines of evolution that might lead to failures of the canisters or radionuclide releases; analyses of the potential rates of release of radionuclides from failed canisters (retention, transport, distribution within repository system and surface) and their potential to give radiation doses to humans, plants, and animals; models and data used in description of evolution of repository system and the development of surface system as pertains to activity releases and dose assessment; a range of qualitative evidence and arguments that complement and support the reliability of the results of the

quantitative analyses; and a comparison of the outcome of these analyses with the safety requirements." See Posiva Oy, "Safety Case for the Disposal of Spent Nuclear Fuel at Olkiluoto: Synthesis 2012," http://www.posiva.fi/files/2987/Posiva_2012 -12web.pdf.

4. Posiva Oy, "Safety Case for the Disposal of Spent Nuclear Fuel at Olkiluoto: Models and Data for the Repository System 2012," http://www.posiva.fi/files/3441 /Posiva_2013-01Part1.pdf.

5. Posiva Oy, "Safety Case for the Disposal of Spent Nuclear Fuel at Olkiluoto: Synthesis 2012."

6. Michael Tomko, "Varieties of Geological Experience: Religion, Body, and Spirit in Tennyson's In Memoriam and Lyell's Principles of Geology," *Victorian Poetry* 42, no. 2 (2004): 119.

7. Posiva Oy, "Safety Case for the Disposal of Spent Nuclear Fuel at Olkiluoto: Synthesis 2012."

8. Anna Weichselbraun, "Constituting the International Nuclear Order: Bureaucratic Objectivity at the IAEA," doctoral dissertation, University of Chicago, Department of Anthropology, 2016.

9. Annelise Riles, *The Network Inside Out* (Ann Arbor: University of Michigan Press, 2000), 22.

10. T. Hjerpe, A. T. K Ikonen, and R. Broed, "Biosphere Assessment Report 2009," Posiva Oy Databank (2009): 14, 19, http://www.posiva.fi/en/databank/biosphere _assessment_report_2009.1867.xhtml#.VcPF6Ra8_dk.

11. Posiva Oy, "Safety Case for the Disposal of Spent Nuclear Fuel at Olkiluoto: Biosphere Assessment 2012," 8, http://www.posiva.fi/files/3195/Posiva_2012-10.pdf.

12. Posiva Oy, " Radionuclide Release and Transport—RNT-2008," 8, http://www .posiva.fi/files/825/Posiva_2008-06_web.pdf.

13. Grasping how these chains work can be tricky. But let's take a stab at it by returning to our previous example of how the Groundwater Flow Model (GFM) was fed into the Radionuclide Transport (RNT), which was then fed into the Biosphere Assessment (BSA). We can approach this chain as if it is a little logic game. From the BSA's perspective *looking backward along the chain* toward the RNT, the RNT appeared as only one *part* that fed into it. This was because, remember, the BSA consumed the RNT as an *input*. Yet from the GFM's perspective *looking forward in the chain* toward the RNT, the RNT appeared as an engulfing *whole* that ingested it as a *part*. This was because, remember, the RNT consumed the GFM as an *input*. Now, let's put these two chain-links together: when the GFM was inputted into the RNT, it helped the RNT produce outputs of its own, which were then fed into the BSA to help make it complete. Finally, the BSA's outputs were input into Finland's nuclear regulator STUK's requirements to help assess the Olkiluoto repository's multimillennial safety.

14. Annelise Riles, "Outputs: The Promises and Perils of Ethnographic Engagement after the Loss of Faith in Transnational Dialogue," *Journal of the Royal Anthropological Institute* 23, no. 51 (2017): 185.

15. There would, for example, always be a time gap between the Radionuclide Transport (RNT) model's and Biosphere Assessment model (BSA)'s submissions: since having a completed RNT was a prerequisite for completing the BSA, delays from the former became delays for the latter too. In practice, though, much of the RNT and BSA were developed simultaneously: since not every aspect of the BSA hinged on the RNT being complete, the BSA experts could work on those parts until they received the inputs from the RNT team needed to finish the rest. The RNT and BSA, then, existed as *before and after* in relation to one another in one sense (as links in a logical chain of inputs/outputs connecting reports), but also *parallel and simultaneous* in relation to one another in another sense (in the living world of Safety Case expert workflows).

16. Roy Wagner, *Symbols that Stand for Themselves* (Chicago: University of Chicago Press, 1986), 3.

17. Marilyn Strathern, "Future Kinship and the Study of Culture," *Futures* 27, no. 4 (1995): 428.

18. Terrence Deacon, *Incomplete Nature: How Mind Emerged from Matter* (New York: W. W. Norton, 2012).

19. Eduardo Kohn, *How Forests Think: Toward an Anthropology Beyond the Human* (Berkeley: University of California Press, 2012), 19–20, 160, 227.

20. Gregory Bateson, *Mind and Nature: A Necessary Unity* (New York: Dutton, 1979), 8.

21. As anthropologist Hirokazu Miyazaki has shown, extensible concepts like "gift" or "arbitrage" can "replicate" themselves across different "spheres of life," eliminating differences as they go, but also serving as means for imagination, speculation, and creative work too. See Hirokazu Miyazaki, "From Sugar Cane to 'Swords': Hope and the Extensibility of the Gift in Fiji," *Journal of the Royal Anthropological Institute* 11, no. 2 (2005): 277–295. See also Hirokazu Miyazaki, *Arbitraging Japan: Traders as Critics of Capitalism* (Berkeley: University of California Press, 2013).

22. Robert Macfarlane, *Underland: A Deep Time Journey* (New York: W. W. Norton, 2019), 15.

23. For more on how ancient conceptual distinctions can become frameworks for deep time reckoning, see Vincent Ialenti, "Adjudicating Deep Time: Revisiting the United States' High-Level Nuclear Waste Repository Project at Yucca Mountain," *Science & Technology Studies* 27, no. 2 (2014).

24. Richard Irvine, "Deep Time: An Anthropological Problem," *Social Anthropology* 22, no. 2 (2014): 162.

25. Annelise Riles, *Documents: Artifacts of Modern Knowledge* (Ann Arbor: University of Michigan Press, 2006).

CHAPTER 3

1. Edmund Burke, *A Philosophical Enquiry into the Origin of Our Ideas of the Sublime and Beautiful* (London: Routledge and Kegan Paul, 1958 [1757]), 39, 72.

2. Marcia Bjornerud, *Timefulness: How Thinking Like a Geologist Can Help Save the World* (Princeton, NJ: Princeton University Press, 2018), 178.

3. Taimi was originally from elsewhere in Europe, but wanted her pseudonym to be "Taimi," a Finnish name.

4. Annelise Riles, "Anthropology, Human Rights, and Legal Knowledge: Culture in the Iron Cage," *American Anthropologist* 108, no. 1 (2006): 52–65.

5. Max Weber, "Science as Vocation," in *The Vocation Lectures*, trans. R. Livingstone, ed. D. Owen and T. Strong (Indianapolis: Hackett Books, 2004 [1917]).

6. Pope Francis, *Laudato Si': On Care for Our Common Home*, encyclical (May 24, 2015).

7. Paul Edwards, *A Vast Machine: Computer Models, Climate Data, and the Politics of Global Warming* (Cambridge, MA: MIT Press, 2003), 221–222.

8. Hirokazu Miyazaki, "The Temporalities of the Market," *American Anthropologist* 105, no. 2 (2003): 255–265.

9. Barbara Adam, *Timescapes of Modernity* (London: Routledge, 1998).

10. E. E. Evans-Pritchard, "Nuer Time-Reckoning," *Africa* 12 (1939).

11. Arie de Geus, *The Living Company* (Boston: Harvard Business School Press, 1997).

12. Michael Sheetz, "Technology Killing Off Corporate America: Average Life Span of Companies Under 20 Years," *CNBC.com*, August 24, 2007, https://www.cnbc .com/2017/08/24/technology-killing-off-corporations-average-lifespan-of-company -under-20-years.html.

CHAPTER 4

1. Vincent Ialenti, "Death and Succession among Finland's Nuclear Waste Experts," *Physics Today* 70, no. 10 (2017).

2. M. Lincoln and L. Bruce, "Toward a Critical Hauntology: Bare Afterlife and the Ghosts of Ba Chúc," *Comparative Studies in Society and History* 57, no. 1 (2015): 191–220.

3. Seneca, *Dialogues and Letters*, trans. R. Campbell (London: Penguin Press, 1997).

4. M. Toth, "Toward a Theory of the Routinization of Charisma," *Rocky Mountain Social Science Journal* 9, no. 2 (1971).

5. Max Weber, *Economy and Society* (Berkeley: University of California, 1978 [1922]).

6. Lamont Lindstrom, "Doctor, Lawyer, Wise Man, Priest: Big-Men and Knowledge in Melanesia," *Man* 19, no. 2 (1984): 291–309.

7. Søren Kierkegaard, *Repetition and Philosophical Crumbs* (Oxford: Oxford University Press, 2009 [1843]).

8. Joel Robbins, "Where in the World Are Values? Exemplarity and Moral Motiva-tion," in *Moral Engines: Exploring the Ethical Drives in Human Life*, ed. C. Mattingly, R. Dyring, M. Louw, and T. Wentzer (New York: Berghahn Books, 2017).

9. This is a reference to anthropologist Kilroy-Marac's ethnographic description of how French military psychiatrist Henri Collomb haunted Dakar, Senegal's Fann

Clinic. For more, see K. Kilroy-Marac, "Speaking with Revenants: Haunting and the Ethnographic Enterprise," *Ethnography* 15, no. 2 (2014): 256, 266, 269.

10. Jacques Derrida, *Specters of Marx: The State of the Debt, the Work of Mourning, and the New International* (New York: Routledge, 1993), 7.

11. Avery Gordon, *Ghostly Matters: Haunting and the Sociological Imagination* (Minneapolis: University of Minnesota Press, 2008), xvi.

12. See Igor Kopytoff, "Ancestors As Elders in Africa," *Africa: Journal of the International African Institute* 41, no. 2 (1971): 138; David Graeber, "Dancing with Corpses Reconsidered: An Interpretation of Famadihana," *American Ethnologist* 22, no. 2 (1995): 258.

13. Friedrich Nietzsche, *The Genealogy of Morals* (Mineola, NY: Dover, 2003 [1887]).

14. International Atomic Energy Agency, "Knowledge Management for Nuclear Industry Operating Organizations," *IAEA-TECDOC-1510* (2006): 56–57, http://www -pub.iaea.org/MTCD/publications/PDF/te_1510_web.pdf.

15. Hugh Gusterson, "Taking RRW Personally," *Bulletin of the Atomic Scientists* 63, no. 4 (2007).

16. Hugh Gusterson, "A Pedagogy of Diminishing Returns: Scientific Involution across Three Generations of Nuclear Weapons Science," *Pedagogy and the Practice of Science: Historical and Contemporary Perspectives*, ed. David Kaiser (Cambridge, MA: MIT Press, 2005), 87.

17. J. Nicholson and R. Corbett, "Key Man Insurance and Market Reaction: A Comment," *Journal of Insurance Issues and Practices* 10, no. 1 (1987).

18. D. Beunza and D. Stark, "The Organization of Responsiveness: Innovation & Recovery in the Trading Rooms of Lower Manhattan," *Socio-Economic Review* 1 (2003).

19. Gregory Bateson, *Steps to an Ecology of Mind* (Chicago: University of Chicago Press, 2000 [1972]), 467.

20. Bruno Latour, "On the Partial Existence of Existing and Non-Existing Objects," in *Biographies of Scientific Objects*, ed. Lorraine Daston (Chicago: Chicago University Press, 2000), 263.

21. Dominic Boyer, *Spirit and System: Media, Intellectuals, and the Dialectic in Modern German Culture* (Chicago: University of Chicago Press, 2005).

22. Charles Perrow, *The Next Catastrophe: Reducing Our Vulnerabilities to Natural, Industrial, and Terrorist Disasters* (Princeton, NJ: Princeton University Press, 2007).

23. Michael Madsen, "The Challenge of Managing Nuclear Knowledge," International Atomic Energy Agency (2014), https://www.iaea.org/newscenter/news /challenge-managing-nuclear-knowledge.

24. Robert Hertz, *Death and the Right Hand* (Glencoe, IL: Free Press, 1960 [1907]).

CONCLUSION

1. Marcia Bjornerud, *Timefulness: How Thinking Like a Geologist Can Help Save the World* (Princeton, NJ: Princeton University Press 2018), 20.

2. David Graeber, "On the Phenomenon of Bullshit Jobs: A Work Rant," *Strike!* (August 2013).

3. David Graeber, "Of Flying Cars and the Declining Rate of Profit," *Baffler* 19 (2012).

4. Richard Nelson, *Make Prayers to the Raven* (Chicago: University of Chicago Press, 1983).

5. Bjornerud, *Timefulness*, 162.

6. Ray Kurzweil, *The Singularity Is Near: When Humans Transcend Biology* (New York: Viking Press, 2005).

7. Frederick Engels, *Socialism: Utopian and Scientific* (Moscow: Progress Publishers, 1970 [1880]).

8. Alan Weisman, *The World without Us* (New York: Thomas Dunne Books/St. Martin's Press, 2007).

9. Lewis Dartnell, *The Knowledge: How to Rebuild Our World from Scratch* (New York: Penguin Books, 2014).

10. Walter Miller Jr., *A Canticle for Leibowitz* (New York: HarperCollins, 1959).

11. Aldo Leopold, *A Sand County Almanac* (New York: Oxford University Press, 1949).

12. Frank Pasquale, ed., *Care for the World: Laudato Si' and Catholic Social Thought in an Era of Climate Crisis* (Cambridge: Cambridge University Press, 2019).

13. Andrew Merrie, "Can Science Fiction Reimagine the Future of Global Development?" *Re.Think* (2017).

14. Rob Nixon, *Slow Violence and the Environmentalism of the Poor* (Cambridge, MA: Harvard University Press, 2011).

15. Simon Nicholson and Sikina Jinnah, *New Earth Politics: Essays from the Anthropocene* (Cambridge, MA: MIT Press, 2016).

16. Kathleen Stewart and Susan Harding, "Bad Endings: American Apocalypsis," *Annual Review of Anthropology* 28 (1999): 286, 291.

17. Richard Irvine, "Deep Time: An Anthropological Problem," *Social Anthropology* 22, no. 2 (2014): 162.

18. Anna Tsing, *The Mushroom at the End of the World: On the Possibility of Life in Capitalist Ruins* (Princeton, NJ: Princeton University Press, 2015): 37.

19. This question is inspired in part by conservation biologist David Ehrenfeld's book on balancing nature, community, and technology, *Becoming Good Ancestors: How We Balance Nature, Community, and Technology* (New York: Oxford University Press, 2008).

20. Gabrielle Hecht, "Interscalar Vehicles for an African Anthropocene: On Waste, Temporality, and Violence," *Cultural Anthropology* 33, no. 1 (2018).

21. Bjornerud, *Timefulness*, 163.

22. Edmund Husserl, *The Phenomenology of Internal Time Consciousness* (Bloomington: Indiana University Press, 1964 [1928]).

23. Henri Bergson, *Time and Free Will* (Mineola, NY: Dover, 1889).

24. David Graeber, *Debt: The First 5,000 Years* (New York: Melville House, 2011).

25. J. G. Ballard, *The Drowned World* (New York: Berkley Books, 1962).

26. Ranjit Singh and Joan Donovan, "The Future Will Be Terrible," *Annual Meeting: Society for Social Studies of Science* (September 4–7, 2019).

27. Paul Tillich, *The Eternal Now* (New York: Scribner, 1963).

28. William Lane Craig, *Time and Eternity: Exploring God's Relationship to Time* (Wheaton, IL: Crossway Books, 2001), 19.

29. "The Rosetta Project: Disk," The Long Now Foundation Library of Human Language, http://rosettaproject.org/disk/concept/ (accessed January 1, 2020).

30. Martin Rees, *On the Future: Prospects for Humanity* (Princeton, NJ: Princeton University Press, 2018), 202.

INDEX